Management for Professionals

More information about this series at http://www.springer.com/series/10101

Sarah Margaretha Jastram • Anna-Maria Schneider
Editors

Sustainable Fashion

Governance and New Management Approaches

Springer

Editors
Sarah Margaretha Jastram
Hamburg School of Business Administration
Hamburg, Germany

Anna-Maria Schneider
Federal Ministry for Economic Cooperation and Development
Berlin, Germany

ISSN 2192-8096 ISSN 2192-810X (electronic)
Management for Professionals
ISBN 978-3-030-08974-0 ISBN 978-3-319-74367-7 (eBook)
https://doi.org/10.1007/978-3-319-74367-7

Cover illustration: eStudio Calamar, Berlin/Figueres

Printed on acid-free paper

This Springer imprint is published by the registered company Springer International Publishing AG part of Springer Nature.
The registered company address is: Gewerbestrasse 11, 6330 Cham, Switzerland

The original version of this book was revised. An erratum to this book can be found at https://doi.org/10.1007/978-3-319-74367-7_15

Contents

Part I
Introduction

Introduction to Sustainable Fashion Governance

Anna-Maria Schneider and Sarah Margaretha Jastram

Fashion is a daily element of everybody's life. We buy fashion items just as any other consumer goods, and we typically judge the clothes we buy based on our assessment of their style, price, and quality. But how are our clothes being made? Many of us do not know much about the production processes and supply chains that are involved in the creation of the T-shirts, trousers, or shoes that we wear every day.

The media keeps reporting about bad working conditions and lacking safety provisions at fashion production sites, about the suppression of trade unions, or about devastating environmental impacts of fashion production and its disposal. Yet, many consumers feel unable to respond to ethical scandals in the fashion industry because the causes of these problems as well as the solutions to them are rather non-transparent or unknown to the public.

This book aims to close this gap. We introduce new governance approaches as well as promising business models and ideas that all share the goals of fostering more decent working conditions in fashion production and of treating the planet more responsibly and sustainably.

By combining perspectives from the governmental sphere, the scientific world, and the business sector, we share current best practice approaches to enable mutual learning as well as co-creation of sustainable solutions to today's challenges in global fashion production.

We will discuss new forms of regulation on the governmental level as well as voluntary private initiatives.

A.-M. Schneider
Federal Ministry for Economic Cooperation and Development, Berlin, Germany

S. M. Jastram (✉)
Department of Strategy and Leadership, Hamburg School of Business Administration,
Dr. Jürgen Meyer Chair of International Business Ethics and Sustainability, Hamburg, Germany
e-mail: sarah.jastram@hsba.de

S. M. Jastram, A.-M. Schneider (eds.), *Sustainable Fashion*, Management for Professionals, https://doi.org/10.1007/978-3-319-74367-7_1

Our core questions are: *What is the current state of sustainable management in the textile and garment industry, and what are successful forms of sustainable fashion governance?*

The textile and garment industry has been known for its serious social and environmental issues for decades. In the 1990s, reports on sweatshops and unacceptable working conditions, including exploitation and sexual harassment, were published. While China and Bangladesh were the first countries to be associated with poor social and environmental standards in fashion production, it is now Myanmar and Ethiopia that have increasingly awakened the interest of the fashion industry due to even cheaper production conditions.

At the same time, reports on poor working conditions in fashion production do not ebb away (ILO 2014) and keep fuelling voluntary engagement in sustainable management approaches by those buyer firms that do not want to contribute (anymore) to the ethical race to the bottom in the fashion industry.

As first reactions, globally sourcing firms have started to implement ethical codes of conduct in their supplier networks. Next, social and environmental audits became common practice to monitor compliance and as part of corporate risk management systems (Schneider and Schwerk 2010). However, Petersen and Krings (2009) highlighted that existing empirical studies do not offer reliable findings about whether codes of conduct actually lead to improvements in suppliers' business practices. According to Li and Zhang (2010), there is a general lack of scientific literature on the determinants of a successful implementation of sustainability in emerging and developing countries as well as on best practice approaches of how to improve working conditions in supplier firms.

Textile value chains are very complex, fragmented over several countries, and often lack transparency (Jastram and Schneider 2015). International frameworks such as the ILO Core Conventions, the Tripartite Declaration of Principles Concerning Multinational Enterprises and Social Policy, the UN Guiding Principles on Business and Human Rights, and the OECD Guidelines for Multinational Enterprises aim to align global business practices. However, effective incentive and enforcement mechanisms are often missing on the international level. Furthermore, national legislation is also often not sufficient to cope with social and environmental challenges related to global value chains and linked international supplier networks. Some governments, however, initiate new national legislation to address negative impacts of international business conduct. In the United Kingdom, for instance, the Modern Slavery Act adopted in 2015 requires bigger companies to create transparency along their supply chains and to report on their efforts to prevent slavery and human trafficking. The French corporate duty of vigilance law was adopted in 2017 and requires the largest French companies to assess and address negative impacts caused by their business activities by publishing annual reports and public vigilance plans. The Child Labour Due Diligence Bill by the Dutch parliament aims to increase due diligence and to combat child labour in global value chains.

Besides this current rise of new mandatory governmental regulation in some European countries, we find a diversified movement towards more sustainable business practices in the world's textile and garment industry.

We are currently observing an increasing variety of private sustainable fashion strategies, initiatives, and instruments as well as public regulations, for which we use the umbrella term Sustainable Fashion Governance in this book.

Sustainable Fashion Governance can be defined as the sum of actors, initiatives, instruments, standard systems, and strategies aiming to force social and environmental sustainable practices in the textiles and garment industry (Jastram and Schneider 2015).

While more and more sustainable fashion governance initiatives arise, the effectiveness and impact of many approaches are contested. Due to the structure of the textile industry, the diversity of business models, and national institutional environments, there is no "one size fits all solution". Large retailers and brands have a different influence on producers than small and medium sized enterprises (SMEs). In consequence, strategies and approaches to improve social and environmental conditions along the supply chain vary substantially.

The transformation of a sector starts with single actors and single steps. Front-runners are the engine for followers and sustainability challenges in global supplier networks will not be solved by a single stakeholder group alone. It takes commitment of various stakeholders to develop and to implement comprehensive and industry-wide solutions in the textile industry. Consequently, the governments of Germany and the Netherlands support sector-wide multi-stakeholder-initiatives (i.e. the Partnership for Sustainable Textiles and the Agreement on a Sustainable Garment and Textile Sector) to improve social and environmental conditions in the textile industry. These initiatives bundle market power and resources of their members to achieve positive impact in production countries.

In this context, this edited volume is an exploration of varied sustainability oriented governance and business approaches. It comprises 12 articles from professional experts and academics. The book provides insights on different spheres of influence and it ushers new perspectives and best practice examples related to the effective implementation of sustainable standards in the fashion industry.

Structured thematically, the first part of the book is dedicated to regulatory frameworks, agreements, and governance approaches to support sustainable development in the textile and garment industry. In the second part, business experts, managers, and entrepreneurs share their experience with the development of new products, business models, and management approaches for sustainable fashion. The following sections will introduce the chapters of the book.

1 First Part

In the first chapter, *Natascha Weisert* explains how the UN Guiding Principles for Business and Human Rights (UNGP), as part of the international regulatory framework, contribute to facilitate a more sustainable fashion industry. The UNGP apply the protect, respect, and remedy framework to clarify roles and responsibilities of states and business as well as other actors to protect human rights in a globalized world. In 2011, the European Commission called for National Action Plans (NAPs) to facilitate the application of the UNGP in the national context. Today, 13 countries have launched a NAP, and more than 20 countries are in the process of developing one.

Under the German Presidency in 2015, the Group of Seven (G7) committed in Elmau to strengthen sustainable practices in global value chains. Among others, the heads of states of G7 agreed to support multi-stakeholder-initiatives to improve the implementation of and compliance with social and environmental standards. Furthermore, production countries will be supported to be part of and to take advantage of responsible global supply chains. Creating a common understanding of due diligence for SMEs will further support the proliferation of sustainable business practices. *Roger Fischer's* article emphasises how an international forum such as G7 creates an international framework towards sustainable global value chains.

Dorothée Baumann-Pauly, Sarah Labowitz and Nate Stein's chapter focusses on a country which is well known in the sustainable fashion debate, mostly as an example of bad production conditions: Bangladesh. The authors analyse the effectiveness of the Bangladesh Accord on Fire and Building Safety and the Alliance for Bangladesh Worker Safety, initiatives which had been established after the fatal catastrophe of Rana Plaza. The chapter argues for an industry transformation in Bangladesh based on Sharing Responsibility, which is an interesting sector-wide approach to sustainability in the textile and garment sector.

2 Second Part

The second part of the book is dedicated to outstanding business approaches in the field of sustainable fashion.

Tchibo aims to reach the peak with its corporate strategy that envisions a 100% sustainable business. The firm's human rights programme shows how companies can align their engagement with existing international frameworks such as the ILO Core Conventions or the UN Guiding Principles on Business and Human Rights. *Nanda Bergstein* demonstrates how human rights can be put into practice by fostering an active dialogue with workers on the supplier level. Tchibo's We-project is acknowledged as a best practice example in the sustainability community.

Turning supplier firms into strategic partners is a successful way to improving sustainability in producing countries and to creating added-value for H&M. *Hendrik Heuermann* points out the rationale behind and the benefits realized through such partner programmes. H&M is going far beyond the traditional audit approach with its Supplier Impact Partnership Programme and joint Sustainability Commitments. The figures presented in the article show the progress made at production sites.

The Otto Group is a big retailer with a huge assortment, thousands of supplier firms, and is confronted with complex challenges when implementing sustainability in the entire supply chain. The article by *Andreas Streubig* gives us interesting insights in Otto Group's management system called 'impACT'. Otto's materiality analysis is based on quantitative and qualitative assessments and complemented by dialogue with internal and external stakeholders. An input-output-calculation model quantifies undesirable environmental and social impacts. This holistic management tool links purchasing and sales data of the Otto Group with specific sustainability aspects.

An entirely holistic approach is pursued by the Swedish fashion label Filippa K, starting with the design phase as the nucleus for sustainable fashion governance. *Elin Larsson* highlights the need to root sustainability in the beginning of the garment life cycle. Filippa K's strategy is to create long lasting clothes in style and quality. The basic idea is not to offer short-dated trends. Timeless pieces and favourite dresses are made to be used season after season. Furthermore, the company increased the number of sustainable fibres to address social issues in the value chain and to decrease any negative environmental impact. The firm's sustainable strategy was enhanced by the first Filippa K second hand store in 2008 which was invented to further prolong the life-cycle of the clothes.

Do good and talk about it? Corporate reporting might not reflect the overall sustainability performance of a company. However, in the academic literature, reporting is an indicator for the engagement of a company. Luxury fashion due to its long-life and high-quality approach has the potential to become a pioneer in sustainable business. *Isabel Laura Ebert's* article is taking a closer look at luxury companies' human rights disclosure. She detects that besides a small number of front-runners there is immense unused potential in the luxury fashion sector to foster more sustainable fashion production.

The MELAWEAR case described by the firm's founder *Henning Siedentopp* is interesting due to its ecological, single country approach. MELAWEAR's supply chain is not fragmented over many different countries, instead the company focuses on one country only—India. The raw material for all products is Fairtrade certified organic cotton. The different processing steps, including dyeing and stitching, are conducted in India and certified by the Global Organic Textile Standard (GOTS). MELAWEAR's business model minimizes negative externalities caused by transportation. Moreover, the firm ensures maximum of value creation for the textiles' country of origin.

Upcycling is another interesting method to prolong the lifecycle of textiles and to contribute to sustainability in the fashion industry. TAUKO is an innovative example of how to use discarded service textiles as raw material for high quality womenswear. The firm's strategy described by *Mila Moisio* focuses on three elements: interesting designs of high quality dresses, refinement of used textiles, and local production. The textiles are coming from the health care, hotel, and gastronomy sectors. Subsequent processing of the dresses takes place within a network of regional partners including suppliers in Finland and Germany.

Besides these innovative business models, consumer decision-making has strong impacts on the pace of the sustainability movement. How can consumers effectively retrace where their clothes are coming from and under which condition they have been made? The Respect-Code is one option to increase transparency in global value chains. *Robin Cornelius* has extensive experience in this field and describes how different fashion production steps from raw material supply to retail can be made retraceable by this instrument. The platform provides production-linked information on participating companies including production facilities, audit results, certificates as well as pictures and contact details. The platform further publishes product-linked indicators, such as CO_2 emissions, water-use footprint, and product-linked certifications and is an interesting tool for consumers and other stakeholders.

Finally, *Samira Iran* is raising the question of whether sustainable fashion needs new forms of consumption and consumer responsibility? The article shows the significant role and power of consumers to reduce negative environmental impacts in all phases of fashion consumption (purchase, usage, and post-usage phase). She emphasizes the need for rethinking our own behaviour and presents creative and innovative alternatives for fashion consumption. Consumers can use second hand clothes instead of new ones or make their own clothes available for the second-hand market. There are different options as part of what Samira Iran calls 'collaborative fashion consumption' such as swapping, borrowing, lending, renting, or leasing clothes, which all provide consumers with opportunities to take over responsibility and to contribute to a more sustainable form of fashion consumption in the future.

In the conclusion, we are summarizing and integrating the key findings of this book on the current state of sustainable fashion governance as well as on current best practice business solutions.

We thank all authors for their valuable inputs to this volume and for sharing their knowledge and expertise on current best practice approaches to make fashion more sustainable. We further thank Zara Berberyan and Johannes Debelius for their support during the final editorial process. Last but not least, we thank our families for their personal support and love.

References

International Labor Organization (ILO) (2014) Wages and working hours in the textiles, clothing, leather and footwear industries. Issues paper for discussion at the Global Dialogue Forum on Wages and Working Hours in the Textiles, Clothing, Leather and Footwear Industries (GDFTCLI). http://www.ilo.org/wcmsp5/groups/public/---ed_dialogue/---sector/documents/publication/wcms_300463.pdf. Accessed 20 Aug 2015

Jastram S, Schneider A-M (2015) Sustainable fashion governance at the example of the partnership for sustainable textiles. uwf Umwelt Wirtschafts Forum 23(4):205–212

Li W, Zhang R (2010) Corporate social responsibility, ownership structure, and political interference: evidence from China. J Bus Ethics 96(4):631–645

Petersen L-E, Krings F (2009) Are ethical codes of conduct toothless tigers for dealing with employment discrimination? J Bus Ethics 85(4):501–514

Schneider A-M, Schwerk A (2010) Corporate social responsibility in Chinese supplier firms. Z Betriebswirt ZfB-Special Issue 1/2010:39–59

Part II
Governance and Regulation

The UN Guiding Principles for Business and Human Rights as a Framework for Action in Global Textile Supply Chains

Natascha Weisert

1 Introduction

Rapidly growing and ever more complex business relationships have been a defining feature of globalization to date. On the positive side, such linkages have led to a significant transfer of skills and technologies, better access to a wider range of products and services, and improved efficiency as well as faster growth. As global supply chains map global production processes across numerous countries at different stages of development and with varying degrees of stability, governance, and respect for the rule of law, they can also serve to trace the negative impacts of globalization, including human rights violations.

In the textile industry, as manufacturers increasingly outsource their labour-intensive production to low cost countries with relatively low sustainability and human rights requirements, such negative impacts can refer to wages below the subsistence level, pervasive overtime work, and a glaring absence of occupational health and safety measures, collective bargaining mechanisms, and unionization. Recent examples with devastating consequences, such as the Rana Plaza tragedy in Bangladesh or the Ali Enterprises factory fire in Pakistan, immediately come to mind, but it is also the continued reports of child and forced labour—be it in cotton production, dye-works or in the finishing stages of textile production—that have led to increasing global attention and outrage (Business and Human Right Resource Center 2017a).

With the aim of addressing the negative human rights impacts in the context of global production processes, the UN Guiding Principles for Business and Human Rights (UNGP) focus on global supply chains to clarify the roles and responsibilities of businesses and other stakeholders (OHCHR 2011). Industries with a strong international orientation or participation, e.g. those relating to large infrastructure

N. Weisert (✉)
Federal Ministry for Economic Cooperation and Development, Bonn, Germany

S. M. Jastram, A.-M. Schneider (eds.), *Sustainable Fashion*, Management for Professionals, https://doi.org/10.1007/978-3-319-74367-7_2

projects (Business and Human Right Resource Center 2017b) or natural resource extraction (Business and Human Right Resource Center 2017c), have long turned to the UNGP for advice on human rights matters. Others, including the textile industry, have only more recently come to embrace the UNGP, possibly fuelled by the previously referenced disasters. In what follows, we therefore explore the provisions of the UNGP to address human rights challenges in global supply chains and discuss, given that they were meant only to provide high-level guidance, through what practical approaches positive impact on the ground could best be obtained.

2 The UNGP as a Global Framework for Business and Human Rights

In 2011, the UNGP were unanimously endorsed by the UN Human Rights Council (UN Human Rights Council 2008; OHCHR 2011). Nowadays, they are widely recognized as the key set of internationally agreed upon norms governing the relationship between states, businesses, and human rights. Besides the approval at the level of nation states and the wide-ranging support from international organizations, business associations and civil society actors, the UNGP are also said to have been "widely taken up by companies as the reference point when implementing human rights considerations into business processes" (CSR Europe et al. 2013; BIAC/ICC/IOE 2011; Oxfam 2013; European Coalition for Corporate Justice 2016). The fact that the UNGP were explicitly taken into account in revising the OECD Guidelines for Multinational Enterprises to include a human rights chapter and in updating the Global Reporting Initiative's (GRI) Sustainability Reporting Guidelines as well as other international corporate social responsibility frameworks further attests to their success (Global Compact Network Germany et al. 2013: 19).

3 From the PRR Framework to the UNGP

The debate surrounding the role of states vs. that of businesses in protecting human rights in a globalized world had, for quite some time, been paralyzed by the seeming impossibility of reconciling legally binding instruments on the one hand with voluntary approaches on the other (Global Compact Network Germany et al. 2013: 15ff). Following the failure of negotiations at the level of the United Nations to create a legally binding human rights instrument to regulate business conduct, Professor John Ruggie, the then appointed UN Special Representative for Business and Human Rights, set out to develop the so-called "Protect, Respect, Remedy" (PRR) Framework that provides the conceptual basis for clarifying the roles and responsibilities of states and business as well as other actors. It comprises three distinct but interrelated pillars:

- *Pillar 1: The State duty to protect* against human rights abuses through appropriate policies or regulation, including abuses potentially caused or contributed to by businesses.
- *Pillar 2: The corporate responsibility to respect* human rights, meaning that businesses are expected to avoid infringing human rights and address negative impacts resulting from or related to their business activities.
- *Pillar 3: Access to remedy*, which refers to the provision of effective grievance mechanisms for victims through states and businesses in case negative impacts on human rights do occur.

Building on this conceptual basis, the UNGP then flesh out the operational implications of the PRR Framework and provide more detailed guidance around the three pillars.

Guiding Principles 1–10 start by describing the state duty to protect human rights. Guiding Principle 1 stipulates, for instance, that "States must protect against human rights abuse within their territory and/or jurisdiction by third parties, including business enterprises", while Guiding Principle 2 establishes the need for states also to require businesses to respect human rights extraterritorially. Guiding Principles 3–10 then outline in what areas and how states should go about fulfilling their duty to protect human rights. One of the key messages therein urges states to "consider a smart mix of measures—national and international, mandatory and voluntary—to foster business respect for human rights" (UNGP and commentary 3 OHCHR 2011).

Guiding Principles 11–24 address the corporate responsibility to respect human rights, outlining specific considerations as to how businesses could meet that responsibility in practice. The Foundational Principles 11–15 first clarify the scope of that responsibility (all enterprises, all internationally recognized human rights, direct activities and business relationships), with Guiding Principles 15–24 then detailing the concrete steps that businesses should undertake to heed their responsibility, including the requirement to undertake human rights due diligence, the centrality of ensuring access to grievance mechanisms, and the importance of putting in place processes that enable the remediation of any negative human rights impacts.

Last but not least, the Access to Remedy pillar with Guiding Principles 25–31 relates to both state and corporate actions to help victims obtain access to remedy for business-related human rights abuses. Based on the recognition of a shared responsibility to provide support for and compensate victims, the respective principles address diverse types of mechanisms, including state and non-state, judicial and non-judicial mechanisms. They also include eight criteria for non-judicial grievance mechanisms established by businesses or partnerships between businesses and other stakeholders, such as civil society actors, to be considered effective (GP 31).

By explicitly distinguishing between the international legal obligation of states to protect human rights and the moral obligation and broad public expectation that businesses—individually and independently—respect human rights, but also bringing them together in one coherent framework, Prof. Ruggie succeeded in reinvigorating the stalled debate.

4 Guiding Businesses' Responsibility to Respect

As already hinted at, the concept of human rights due diligence lies at the heart of guiding businesses' responsibility to respect human rights. More specifically, the UNGP's second pillar outlines five essential steps businesses should undertake, including making a policy commitment (UNGP 16); undertaking an assessment of human rights-related risks and (potential) human rights impacts; broadly integrating this understanding and these risks into all relevant business processes and levels; and establishing processes to track the effectiveness of actions undertaken to address human rights issues and to communicate results (UNGP 17–21). In terms of scope, the UNGP require businesses to consider (potentially) adverse impacts throughout their entire supply chains with each and every due diligence step (UNGP 13; OHCHR 2011).

5 Policy Commitment

As a first step in fulfilling their responsibility to respect human rights, UNGP 16 expect businesses to issue a high-level public commitment to respect human rights, be it in the form of a specific human rights policy or as part of other existing policies or relevant codes of conduct. Such a commitment should, at a minimum, specify the most critical human rights issues a business is facing and how these are reflected in its corporate policies and processes, including what is expected of its business partners and other entities linked to its operations, products, or services. See UNGP 16 and commentary (OHCHR 2011) and (Langlois 2016) for further guidance on what a company's human rights policy should contain.

6 Human Rights Due Diligence

UNGP 17 states that "in order to identify, prevent, mitigate and account for how they address their adverse human rights impacts, business enterprises should carry out human rights due diligence" (UNGP 17; UNGP 18; OHCHR 2011), which, according to UNGP 18, requires businesses to proactively "identify and assess any actual and potential adverse human rights impacts with which they may be involved either through their own activities or as a result of their business relationships" (UNGP 18; OHCHR 2011). In practice, this means that businesses have to decide whether it is sufficient to carry out a broader risk assessment of how their overall business activity might impact human rights or whether their specific operations might warrant undertaking an in-depth human rights impact assessment.

The main criterion advanced by the UNGP is the degree of riskiness that a potentially severe human rights impact will in fact occur, taking into account aspects such as the size or dimension of the business venture, the country or sector of operation, a specific product or service as well as other contextual factors (UNGP 17 b and c; OHCHR 2011). UNGP 17 acknowledges that a supply chain may in fact contain so many different entities that a reasonable approach to human rights due diligence in such a context would warrant prioritization of those areas for assessment (operations, products, services, location), where the risk of adverse human rights impacts is highest (UNGP 17; OHCHR 2011).

When it comes to integrating the findings from the risk analyses or impact assessments into business policies or processes, Guiding Principle 19 advocates not only for appropriate processes to be put in place, but also for them to be internally aligned and continually reviewed to match the broader and continuously changing human rights context. This is because adverse human rights impacts can only be avoided, mitigated or effectively dealt with if the analyses or assessments undertaken count on full internal collaboration and if the results are fed back into action in both a horizontal—across departments and operational units—and a vertical manner—from the leadership down to the individual employees.

UNGP 19 also touches on the increasing complexity of dealing with the responsibility to respect in a supply chain context and sets out to provide practical guidance on appropriate business conduct in the context of an adverse or potentially adverse human rights impact, supplementing the decision tree thus far with two further criteria. The first one has to do with the degree to which an enterprise is actually linked to the (potentially) adverse human rights impact. More specifically, the UNGP distinguish between three levels of involvement: a business can cause adverse human rights impacts through its own activities; a business can contribute to such an impact through its own activities; or it can be directly linked to an adverse impact through its business relationships.

In case of an enterprise's activities being a direct cause or contributing to an adverse human rights impact, immediate action to prevent or stop the impact is in order. Where an adverse human rights impact occurs by way of existing business relationships, UNGP 19 stipulates that appropriate action will depend on a number of factors such as "the enterprise's leverage over the entity concerned", and further, "how crucial the relationship is to the enterprise, the severity of the abuse, and whether terminating the relationship with the entity itself would have adverse human rights consequences" (Commentary to UNGP 19 OHCHR 2011).

The "amount of leverage" or the possibility to exercise it is then the second criterion offered by the UNGP for a business to decide on how to address adverse human rights impacts along its supply chain.[1] If there is leverage, meaning that a business can "effect change in the wrongful practices of the party that is causing or contributing to the impact", then, according to UNGP 19, it should be applied. Otherwise, an enterprise could, for one thing, try to increase its leverage, possibly by providing more support to its suppliers through capacity building and/or by fostering

[1] In other words, impacts that it may have contributed to or been directly linked with (see UNGP 19 and commentary).

partnerships that help to further support or solidify the business relationships along the supply chain. For another, it could attempt to sever the relationship through which the human rights impact has occurred, provided that this is possible and does not bring with it other unwanted and negative human rights impacts (UNGP 19 and commentary OHCHR 2011).

In order to carry out human rights due diligence, the fifth step a business has to undertake is to effectively monitor and track the results (UNGP 20) and communicate (UNGP 21) its commitment, analysis, and response to adverse human rights impacts. In order to improve due diligence processes and to adapt to changing (human rights) contexts, all the steps should be repeated in a circular manner, with the insights and results generated by one round feeding into and powering the next.

Beyond that, UNGP 22 provides a way forward in those instances, where businesses have caused or contributed to an adverse human rights impact that they were not able to prevent or foresee. Businesses should then "provide for or cooperate in their remediation through legitimate processes". They should also establish or participate in operational-level grievance mechanisms for those who may be negatively impacted (UNGP 22; OHCHR 2011). UNGP 23 and 24 finish off the discussion by reminding businesses that their human rights responsibility applies anywhere they operate and that they may have to prioritize among alternative courses of action in case of simultaneously occurring human rights impacts (UNGP 23 and 24 as well as the accompanying commentary OHCHR 2011).

7 Mechanisms to Assist UNGP Implementation

Over the past years, the UNGP have stimulated a significant amount of corporate action. Based on the human rights due diligence elements outlined above, businesses have, often in cooperation with other actors, such as specialized consulting firms and expert advisers, established internal processes to help them analyse and monitor their human rights risks and engage in preventive and/or remedial action. Discussions at the annual UN Forum on Business and Human Rights are a clear testimony to these efforts (OHCHR 2017; O'Brien and Dhanarajan 2015; Shift 2012).

Nevertheless, whether recourse to the UNGP has indeed translated into positive human rights-related impact along global supply chains has repeatedly been called into question (Lukas 2011: 7–8; Ruggie 2010). This is because the UNGP, as high-level principles, deliberately leave many critical questions unanswered, such as what the scope of adequate human rights due diligence and risk assessments should be, how the concept of "leverage" should be applied and what exactly constitutes "significant" human rights risks or "serious" human rights abuse. They should be clarified through relevant implementation mechanisms or processes: to overcome conceptual difficulties, practical testing is required; to deal with implementation-related problems, relevant capacities need to be built.

It is also unlikely that an individual enterprise could master this task alone. Some have argued that human rights due diligence could be such a daunting task that businesses might resort to a very superficial analysis or could even be discouraged in the first place from identifying human rights risks and related impacts along their supply chains (Thorsen and Andreasen 2011: 11). Dialogue, exchange, and partnership processes are thus likely to be critical elements in facilitating practical application, further refining the proposed concepts through action-oriented discussion, verification and renewed practical application, all amounting to a comprehensive continuous improvement exercise. They would also help in keeping costs down and motivation (to engage) up.

To assess whether the UNGP provide a meaningful framework for the responsibility to respect human rights along global supply chains, we will therefore turn to three mechanisms or processes through which their clarification and implementation have been pursued in practice and seek to assess the results that these mechanisms have produced. They have, in turn, been commonly favoured by civil society organizations, corporations and business associations, and governments—often in combination with some or all of the other stakeholders.

8 Sustainability and Human Rights Reporting

In the debate around business and human rights, civil society actors, and increasingly also consumers, have been at the forefront of demanding greater corporate transparency. In order to credibly report on their human rights risks and impacts, the reasoning goes that reporting must be based on a solid human rights due diligence process. To facilitate this, civil society organizations have also been supporting efforts to standardize sustainability indicators and to assess and benchmark corporate sustainability reports via relevant initiatives. Launched in cooperation with the United Nations, the Global Reporting Initiative (GRI) has, for instance, made a significant contribution to harmonizing and standardizing sustainability reporting, including by updating its reporting framework following UNGP adoption (Global Reporting 2017; UNGP Reporting Framework 2017).

In response to these demands, business interest in reporting has grown strongly over the past years, especially of large, internationally oriented companies. In 2015, for example, more than 80% of companies listed on the "Standard & Poor's 500" index published sustainability reports, up from a mere 20% in 2011 (Environmental Leader 2016). While these developments should certainly be welcomed, the voluntary nature of sustainability reporting has frequently been criticized.[2] Instead, the advanced proposal is for reporting—or even for human rights due diligence—to be a legal or mandatory requirement, thus raising the pressure on companies to seek

[2]And, so the argument goes, voluntary reporting implies impunity in case reporting is not undertaken or perhaps problematic aspects are overlooked and actual human rights or other abuses occur.

investment to properly and comprehensively analyse and actively manage their human rights risks.

9 Practice-Oriented Business Networks and Standard Initiatives

Businesses have, in the context of the broader corporate sustainability discussion, increasingly sought out collaborative approaches to apply and ensure adherence to sustainability and human rights standards in their supply chains. In these initiatives, companies typically define, often together with other stakeholders, private standards that are then implemented within the company (code of conduct along the supply chain), in the industry (industry or sector initiatives), or across a broad range of private sector actors (e.g. corporate citizenship initiatives or CSR networks) (OECD 2009: 237ff). Over time, their interest in engaging in such initiatives has accelerated, not least through the work of Prof. Ruggie.

> Over the last two decades, voluntary standard-setting initiatives have proliferated, becoming one of the most popular tools for addressing the human rights impact of global business activities. (...) Indeed, over 180 standard-setting initiatives that address social, ethical, or environmental issues have now been mapped (Evans and Winstanley 2015).

What real impact corporate sustainability initiative for the protection of human rights can really produce along global supply chains, however, has not been clear. Voluntary learning fora and CSR initiatives addressing the private sector as a whole can certainly raise awareness of critical sustainability aspects and help to kick start timely and relevant pilot initiatives, but they usually lack appropriate verification or exclusion mechanisms to get their members to establish far-reaching due diligence procedures. Given their broad, cross-sectoral orientation, they are also unlikely to include the specific range of stakeholders required to elicit a strong commitment to respect human rights from any particular member company.

10 National Action Plans on Business and Human Rights

Based on the widespread endorsement of the UNGP and in recognition of their high-level guidance nature, the European Commission called in 2011 for the drawing up of National Action Plans (NAPs) to localize the UNGP and facilitate their application in the national context. The UN Human Rights Council soon followed suit in 2014 (European Commission 2011; UN Human Rights Council 2014).

Accordingly, a NAP should aim at affirming "the UNGP's essential tenet that human rights apply within the business sector and indicates a political commitment to bring domestic laws, policies, and practices into alignment with this norm" (O'Brien et al. 2016: 2). Such a process should include the determination of the country or context-specific smart mix of voluntary and mandatory measures, e.g. by clarifying and broadly communicating human rights-related expectations to businesses, such as by instructing embassies and consulates accordingly, but also by, for instance, developing appropriate regulations and adopting laws that make human rights due diligence a requirement to qualify for public procurement or foreign trade promotion measures (Bauer 2014).

To date, 13 countries have launched a NAP, including the UK, the Netherlands, the US, Germany, and Colombia, and more than 20 countries are in the process of developing one, including France and Japan (National Action Plans 2017). To ensure that NAPs lead to positive human rights impact and pave the way for effective implementation, a number of success criteria have been laid out. In particular, it has been advocated to ensure broad ownership within government and across different societal actors; to develop measures on the basis of a gap analysis (or so-called baseline assessment); and to establish a sound monitoring process with clear objectives, deadlines and assessment criteria (De Felice and Graf 2015).

How relevant have these mechanisms been in practice when it comes to positive human rights impact along global value chains? We now turn to the current experiences and evidence associated with human rights reporting, standard initiatives, and NAP processes.

11 Sustainability and Human Rights Reporting

As far as mandatory corporate reporting and human rights due diligence are concerned, there are undoubtedly instances where this has had a pronounced impact.

> The US Dodd-Frank Consumer Protection Act of 2010 Section 1502 provision on conflict minerals is a case in point. The provision's requirement that companies disclose their use of conflict minerals (...) has forced listed companies to take steps to avoid contributing to the problem, which they had all but ignored prior to the passage of the law (Bauer 2014).

Large corporations such as Intel and Apple have, consequently, developed and certified conflict-free products.

In a similar spirit, the transposition of the "EU Directive on non-financial and diversity-related information by certain large companies and groups" into national law by the end of 2016 was deemed by many an opportunity to oblige large companies to engage in sustainability reporting on the basis of a comprehensive due diligence process. NAP processes in several European countries, e.g. in Germany, sought to leverage the transposition to establish a comprehensive business and human rights framework, including a meaningful smart mix of voluntary and mandatory elements.

The resulting national laws and their potential results to be obtained differ widely however. For example, in Germany, the law affects only about 550 companies, while more than 1000 are addressed in Denmark or Sweden, i.e. much smaller countries. In addition, the national laws use different risk definitions, include more or less explicit requirements to account for business relationships along global supply chains, and prescribe very different audit procedures for reports as well as fines. Certainly worth mentioning is the French Implementation Act, which has come into effect on 27 March 2017. While only applying to just under 100 French companies, it requires reporting on all elements of due diligence and envisages high fines of up to 30 million euros in the case of non-compliance. Furthermore, the UK Modern Slavery Act has committed some 12,000–17,000 large companies to report on how they counteract human trafficking and slavery along their supply chains. Thus far however, only few companies have published a Modern Slavery and Human Trafficking Statement, and state tracking overall appears weak.[3]

Overall, given that the more rigorous reporting requirements are still quite young, it is difficult to draw any firm conclusions on their impact along global supply chains at this stage. Making reporting a legal requirement can certainly provide an important signal for businesses to review and invest in their due diligence processes. A legal requirement on its own is unlikely, however, to be sufficient. To yield demonstrable results and track the impact on the ground, effective enforcement and participatory implementation processes are needed.

12 Practice-Oriented Business Networks and Standard Initiatives

Witnessing the proliferation of private standard initiatives and increasing interest in human rights aspects, such initiatives can certainly provide a relevant framework for the establishment of human rights due diligence procedures. In practice, however, their implementation has rarely yielded much positive impact beyond first tier suppliers, where production processes are more difficult to track, implementation capacities are weak, and human rights risks are more pronounced. They also seem to have a tendency to focus on those human rights that can be more easily addressed, e.g. on working hours and health and safety, as well as on particularly exposed sectors and regions, which means that significant human rights risks can go unchecked (Thorsen and Andreasen 2011: 8).

Sector-specific multi-stakeholder initiatives nevertheless seem to be able to yield effective human rights protection along global supply chains. In particular, due to the involvement of a critical mass of industry-specific companies, governmental members, civil society actors, and trade unions, etc., these largely voluntary initiatives can generate sufficient pressure to function as quasi-binding control mechanisms for all members to heed their provisions. An evidence-based analysis by Baumann-Pauly et al. (2016) does, for instance, provide a very positive assessment of the Fair Labour Association (FLA) and the Global Network Initiative (GNI) with regard to

[3]The UK example refers to an earlier example, which is not related to the transposition of the EU directive.

their reach, the effectiveness of their standard requirements, and their enforcement down to the level of member companies' lower tier suppliers; Thorsen and Andreasen (2011) advance similar results to argue in favour of their so-called third generation responsible supply chain management model, which is, in essence, sector-specific and draws on the participation of a broad range of actors.

13 National Action Plans on Business and Human Rights

An earlier analysis of existing NAPs, based on the success criteria mentioned above, detects much room for improvement in yielding tangible results. No country seems to have conducted a baseline assessment; targets and implementation criteria are missing. While a later analysis points to numerous improvements such as wider stakeholder participation or an increasing number of baseline assessments, a proper smart mix as well as monitoring and accountability mechanisms to support NAP implementation remain absent (De Felice and Graf 2015: 65).

It thus appears that NAP processes as such are also not a guarantor for effective national implementation, including along global value chains. The hope voiced by various authors is for NAPs to become more effective over time. Supported by a global dialogue as well as monitoring and capacity building measures via the United Nations (Working Group on Business and Human Rights), NAP processes are expected to build on successful experiences through continuous review and experimentation (De Felice and Graf 2015: 65, O'Brien et al. 2016: 15).

14 Towards a Comprehensive Business and Human Rights Ecosystem

The adoption of the UNGP in 2011 has fuelled a global dialogue on the roles and responsibilities of states and business in upholding human rights in a globalized world. As far as businesses are concerned, the UNGP provides a detailed discussion of the responsibility to respect, but by virtue of their being high-level principles, more advanced concepts, such as the various steps required to undertake a proper human rights due diligence, need to be interpreted and applied to the local context through practical application, peer pressure, and peer learning. We have looked at several mechanisms that aim to promote and incentivize such contextualization, with a focus on impact along global supply chains.

Like the UNGP, many of these mechanisms are still very young so that a comprehensive assessment of their relevance is difficult to undertake at this stage, especially as regards human rights reporting and NAP processes. The only clear and clearly positive evidence available relates to the experience of sector-specific multi-stakeholder initiatives, as these seem to be able to assemble the right kind of pressure and peer group to mandate, effectively implement, and monitor human rights-related actions along global supply chains.

Such initiatives are nevertheless unlikely to come about just by themselves. They need strong policy signals and incentives and rigorous frameworks to compare their effectiveness and results, as well as relevant platforms to facilitate exchange, benchmarking across sectors, and an alignment of a broad range of business and human rights initiatives.

This is exactly where all the implementation mechanisms could complement and reinforce each other. Reporting frameworks provide guidance on specific results and facilitate the comparison of initiatives and their results. Mandatory reporting requirements emit strong policy signals. Sector-specific initiatives create, apply, and monitor measures to protect human rights beyond individual supply chains. And NAP processes provide a platform "for the critical review of existing regulatory approaches on business and human rights, as well as for dialogue about policy innovations" and the broader political, economic, legal and regulatory factors influencing the success or failure of such mechanisms in practice (O'Brien et al. 2016: 15–16).

While the different stakeholder approaches are extremely important in their own right and respond to the roles and responsibilities laid out by the PRR framework, their full power can be unleashed only by unifying the pillars and building on multi-stakeholder processes in a comprehensive ecosystem for protecting human rights in a context of global business activity.

Disclaimer

The views and opinions expressed in this article are those of the author. They do not purport to reflect the opinions or views of the German government or any of its agencies. Designations employed and assumptions made within the analysis or explanations provided are also not reflective of the position of any German government entity.

References

Bauer J (2014) Where do national action plans on business & human rights belong in the corporate sustainability movement? In: CSRwire Talkback blog of 17 March 2014. http://www.csrwire.com/blog/posts/1262-where-do-national-action-plans-on-business-human-rights-belong-in-the-corporate-sustainability-movement. Accessed 15 June 2017

Baumann-Pauly D et al (2016) Industry-specific multi-stakeholder initiatives that govern corporate human rights standards – legitimacy assessment of the fair labor association and the global network initiative. J Bus Ethics 140(3):1–17

BIAC, ICC and IOE (2011) Joint IOE-ICC-BIAC comments on the draft guiding principles on business and human rights statement of 26 January 2011. https://business-humanrights.org/sites/default/files/media/documents/ruggie/ioe-icc-biac-comments-on-guiding-principles-26-jan-2011.pdf. Accessed 15 June 2017

Business and Human Right Resource Center (2017a) https://business-humanrights.org. Accessed 15 June 2017

Business and Human Right Resource Center (2017b) https://business-humanrights.org/en/guatemala-two-deaths-in-recent-accident-highlight-history-of-human-rights-violations-linked-to-the-world-bank%C2%B4s-chixoy-dam. Accessed 15 June 2017

Business and Human Right Resource Center (2017c) https://business-humanrights.org/en/human-rights-impact-assessment-of-tampakan-copper-gold-project-mindanao-philippines. Accessed 15 June 2017

CSR Europe et al. (2013) Putting the ruggie-framework and the guiding principles into practice. Report on the business and human rights workshop. http://www.econsense.de/sites/all/files/Documtenation%20Workshop%202013_0.pdf. Accessed 15 June 2017

De Felice D, Graf A (2015) The potential of national action plans to implement human rights norms: an early assessment with respect to the UN guiding principles on business and human right. J Hum Rights Pract 7(1):40–71

Environmental Leader (2016) 81% of S&P 500 companies published sustainability reports in 2015. https://www.environmentalleader.com/2016/06/81-of-sp-500-companies-published-sustainability-reports-in-2015. Accessed 15 June 2017

European Coalition for Corporate Justice (2016) 5 years of UNGPs: 5 business & human rights issues to focus on. http://corporatejustice.org/news/135-5-years-of-ungps-5-business-human-rights-issues-to-focus-on. Accessed 15 June 2017

European Commission (2011) A renewed EU strategy 2011-14 for corporate social responsibility. http://eur-lex.europa.eu/legal-content/EN/TXT/PDF/?uri=CELEX:52011DC0681&from=EN. Accessed 15 June 2017

Evans A, Winstanley S (2015) Asking the basic questions: are voluntary standard-setting initiatives protecting human rights? In: UN forum series (Blog). http://blogs.lse.ac.uk/businesshumanrights/2015/11/20/un-forum-series-voluntary-standard-setting-initiatives-protecting-human-rights/. Accessed 15 June 2017

Global Compact Network Germany et al (2013) Respecting human rights – an introductory guide for business. https://www.globalcompact.de/wAssets/docs/Menschenrechte/Publikationen/respecting_human_rights-an_introductory_guide_for_business.pdf. Accessed 15 June 2017

Global Reporting (2017) https://www.globalreporting.org/resourcelibrary/GRI-UNGP_LinkageDoc.pdf. Accessed 15 June 2017

Langlois M (2016) Reporting trends and insights: who's responsible and accountable for addressing human rights risks within the company? http://www.ungpreporting.org/reporting-insights-trends-responsible-accountable-human-rights-risks-company. Accessed 15 June 2017

Lukas K (2011) Human rights in the supply chain: influence and accountability. In: Mares R (ed) The UN guiding principles on business and human rights: foundations and implementation. Nijhoff, Boston

National Action Plans (2017) http://www.ohchr.org/EN/Issues/Business/Pages/NationalActionPlans.aspx. Accessed 15 June 2017

O'Brien CM and Dhanarajan S (2015) The corporate responsibility to respect human rights: a status review. National University of Singapore Law, Working Paper 2015/005. http://law.nus.edu.sg/wps/pdfs/005_2015_Sumithra%20Dhanarajan.pdf. Accessed 15 June 2017

O'Brien CM, Mehra A, Blackwell S, Poulsen-Hansen CB (2016) National action plans: current status and future prospects for a new business and human rights governance tool. Bus Hum Rights J 1(1):117–126. https://doi.org/10.1017/bhj.2015.14. (online version has been updated in 2016 and shows different page numbers, see https://papers.ssrn.com/sol3/papers.cfm?abstract_id=2627568)

Office of the High Commissioner of Human Rights (OHCHR) (2011) Human rights and transnational corporations and other business enterprises, Resolution A/HRC/RES/17/4. https://documents-dds-ny.un.org/doc/RESOLUTION/GEN/G11/144/71/PDF/G1114471.pdf?OpenElement. Accessed 19 June 2017

Office of the High Commissioner of Human Rights (OHCHR) (2017) Forum on business and human rights. http://www.ohchr.org/EN/Issues/Business/Forum/Pages/ForumonBusinessandHumanRights.aspx. Accessed 15 June 2017

Organisation for Economic Cooperation and Development (OECD) (2009) Overview of selected initiatives and instruments relevant to corporate social responsibility. In: OECD (ed) Annual report on the OECD guidelines for multinational enterprises 2008. Employment and industrial relations, Paris, pp 235–260

Oxfam (2013) Business and human rights – an Oxfam perspective on the UN guiding principles. https://www.oxfam.org/sites/www.oxfam.org/files/tb-business-human-rights-oxfam-perspective-un-guiding-principles-130613-en.pdf. Accessed 15 June 2017

Ruggie J (2010) The corporate responsibility to respect human rights in supply chains. In: 10th OECD roundtable on corporate responsibility. https://www.oecd.org/corporate/mne/45535896.pdf. Accessed 15 June 2017

Shift (2012) Respecting human rights through global supply chains. In: Workshop report no. 02. http://www.shiftproject.org/media/resources/docs/Shift_UNGPssupplychain2012.pdf. Accessed 15 June 2017

Thorsen S, Andreasen S (2011) Remodelling responsible supply chain management: the corporate responsibility to respect human rights in supply chain relationships. In: Mares R (ed) The UN guiding principles on business and human rights: foundations and implementation. Nijhoff, Boston

UN Human Rights Council (2008) Protect, respect and remedy: a framework for business and human rights: report of the special representative of the secretary-general on the issue of human rights and transnational corporations and other business enterprises, report prepared by John Ruggie, A/HRC/8/5, 07 April 2008

UN Human Rights Council (2014) Human rights and transnational corporations and other business enterprise, resolution A/HRC/RES/26/22, 15 July 2014

UNGP Reporting Framework (2017) http://www.ungpreporting.org. Accessed 15 June 2017

UNGP United Nations Guiding Principles on Business and Human Rights (2011) Implementing the United Nations "Protect, respect and remedy" framework. http://www.ohchr.org/Documents/Publications/GuidingPrinciplesBusinessHR_EN.pdf. Accessed 15 June 2017

Towards Sustainable Global Value Chains: The Role of the Group of Seven

Roger A. Fischer

At the Elmau summit in June 2015, heads of state and government of the Group of Seven (G7) together committed to undertake a number of policy measures to strengthen sustainable practices in global value chains (G7 (Elmau) Summit Leaders' Declaration 2015: 6–7). This essay sets out the reasons that spoke for joint G7 action towards that goal and gives an account of what the G7 agreed upon in Elmau. It describes some features of the process leading up to the summit, attempts to draw a few lessons and then closes with some open but perhaps relevant questions concerning the future of the agenda for sustainable global value chains.

1 A Case for Action

Why would governments engage in sustainable global value chains? It is a business issue, seemingly different from the typical macroeconomic conversations held by leaders and ministers. The case can be (and indeed has been) made from three different angles: Standards, development, and trade.

1.1 Implementing Social and Environmental Standards

There is a large set of international rules governing production and trade. They reflect a broad consensus among states and many multinational enterprises on how the interface between business and its social and environmental externalities should

R. A. Fischer (✉)
Federal Ministry for Economic Cooperation and Development, Berlin, Germany
e-mail: roger.fischer@bmz.bund.de

S. M. Jastram, A.-M. Schneider (eds.), *Sustainable Fashion*, Management for Professionals, https://doi.org/10.1007/978-3-319-74367-7_3

eventually look like. But these rules are not self-implementing. In developing producer countries particularly, gaps persist between standards and reality. In the absence of effective enforcement, practitioners have often turned to those who benefit when rules are applied. In the case of social and environmental standards, two groups of people stand to benefit: First, workers and citizens in developing producer countries who suffer from infringements and need effective remedies. And second, people in consumer countries who ask for compliant goods (for the growing demand for sustainably produced goods see The Nielsen Company (2015), and for the growing market share of sustainable commodities (coffee, cocoa, palm oil, tea, cotton, bananas, sugar and soybeans) see Terazono (2015)). Using the agency of interested parties can be a powerful tool for rule implementation—witness the class actions broadly on offer in the US legal system (Hay 2002: 178–182) and the European Court of Justice's case law on the direct effect of unimplemented EU directives (Chalmers et al. 2010: 268). It is a small step to think about what this approach can do for international social and environmental norms.

The need for better implementation is particularly evident in the case of core labor standards. Global value chains operate like single integrated factories (Taglioni and Winkler 2016: 11–12). They make different production factors work together for one business purpose. What traditional factories deliver by placing all factors in the same location, new global value chains replicate by sophisticated organization and transport across continents. The concept is the same. But workers in those new "factories" do not make a unified workforce. Unlike the increasingly seamless production process, they remain divided by various effective levels of protection. Every worker is left with the labor legislation of his country and the legal relationships within his company. Variations between companies are important because many sourcing relationships are contractual in nature (Taglioni and Winkler 2016: 12).

1.2 Achieving Development Gains

From a development angle, decent work and sustainable production are not just norms. They are also goals ((See: Transforming Our World 2015), Goals 8 (decent work) and 12 (sustainable consumption and production)), reached as a result of long uncertain processes. Global value chains are a point in case. They can turn into powerful vectors towards better employment, more efficient use of resources and skills enhancement in developing producer countries. But these gains are by no means automatic. Access to global value chains is necessary but not sufficient to trigger them (Taglioni and Winkler 2016: 199).

Development is about taking specific opportunities inherent in larger trends beyond anyone's control. The unprecedented increase in all forms of international division of labor over the last 20 years (Taglioni and Winkler 2016: 16) is clearly among those trends. Trade through global value chains enables development because three factors come together: First, there is a business case for spreading production across borders (Fally and Hillberry 2015). For businesses in rich countries, outsourcing tasks to firms abroad can be a way to escape high ("intra-firm") coordination costs at home and capture lower ("inter-firm") transaction costs elsewhere, using the strong legal infrastructure in your home market and the demand power you can project through your value chain. For businesses in poorer countries, positioning yourself as a supplier can turn your low ("intra-firm") coordination costs and your resulting larger firm scopes into competitive edges, making up for the high transaction cost environment in your country. Second, more open markets and improved trade infrastructure added opportunities to actually operate on those business cases. Third, contemporary management thinking has explored the possibilities of distributing the tasks of production to different specialized business units ("second unbundling"; Taglioni and Winkler 2016: 16). This is an opening for developing countries in particular. They can now enter production without necessarily having the wherewithal to master the entire process (Taglioni and Winkler 2016: 13).

The issue here is to understand what enabled some emerging economies to consistently translate their participation in global value chains into tangible and growing development gains—and then use those lessons to make a difference in other developing countries. While many questions in this context remain open, responsible business conduct certainly plays a role. We used a "factory" simile to illustrate the case for implementing core labor standards throughout global value chains. A related case can be made in development terms. Global value chains can drive development because they convey knowledge; often from "upstream" to "downstream" players[1] (Taglioni and Winkler 2016: 28f). This is why, depending on the sector, (Taglioni and Winkler 2016: 21) "moving upstream" can be used as one very imperfect proxy for upward graduation through global value chains. The business purpose of knowledge conveyed through global value chains is to control production.[2] With control comes responsibility. It is therefore fair to make the normative case that no business leading a global value chain can credibly dissociate itself from the development effects—positive and negative—of that chain. Whoever is in control has to answer for what they could have done. But first movers on responsible business conduct risk to be set back against competition without some basic level of coordination. The G7 provides a platform to offer that coordination and support for responsible business conduct without unduly distorting competition.

1.3 Building Support for Trade

Finally, there is a trade policy angle. Each step towards more openness and integration in global trade now seems much harder than the ones before. What we have

[1] "Upstream" stands for distance; "Downstream" for closeness to final demand.

[2] Think of product specifications, design, intellectual property, management process.

achieved in terms of lower tariffs and fewer restrictions in hindsight feels like easy gains when compared to the "deep integration" and "behind the border" complexities of today's trade negotiations. They just absorb more political capital. It is one thing to change the price environment for domestic producers by abolishing a quota system. It is quite another to invite foreign investors to become domestic producers themselves. This is part of the reason why more openness to trade is not anymore immediately seen as something worth having. The case needs to be made—and not in economic terms only. Witness the way projects for new plurilateral agreements are debated in OECD democracies. Today consumers question where goods come from, if they are safe and how they were made. Workers expect governments to manage the trade-offs between employment goals at home and abroad. One way of answering these concerns is to show that open trade is not pursued as an end in itself, but as one enabler for shared prosperity. Positive measures to enhance the implementation of social and environmental rules and to draw sustainable development gains from global value chains can carry this message.

Up to a point, the three angles converge. Indeed, the Elmau commitments for sustainable global value chains can arguably be seen as a coherent, "one agenda"[3] style programme of measures, taking place in various policy areas but all serving the same goal. However, at least notionally, the three angles still pull in different directions. The following sections will grapple with some of these tensions.

2 What the Group of Seven (G7) Did

2.1 The Elmau Outcome

The package of G7 outcomes reached in 2015 on the issue of sustainability in global value chains has three elements:

First, the Elmau Leaders' Declaration, where the G7 specifically commit: (1) to increase their support to help small and medium enterprises (SME) develop a common understanding of due diligence and responsible supply chain management, (2) to strengthen multi-stakeholder initiatives for sustainable global value chains in G7 countries and beyond, (3) to support partner countries in taking advantage of responsible global supply chains and (4) to strengthen mechanisms for access to remedies under the system of National Contact Points for the OECD Guidelines for Multinational Enterprises. Note that in terms of the structure of the Leaders' Declaration, these commitments are not part of the development section but immediately follow the paragraph on trade.

[3] A term associated with the G8 summit in Lough Erne, where measures on tax, transparency and trade were combined in this way.

Second, the G7 will use agreed-upon indicators and data sources to measure and publicly report on progress against these commitments (G7 Ise Shima Progress Report 2016: 32).[4] And third, the Joint Declaration of G7 Labor and Development Ministers "Action for Fair Production", issued October 13th 2015, set out steps to implement the Elmau commitments.

Before 2015, the G7 and the G8[5] had very often agreed on joint positions concerning trade and development, and once on social standards. Global environmental issues have also been early features of the G7/G8 agenda (G7 Elmau Progress Report 2015: 3). But those positions were normally taken one by one, not in conjunction. A commitment to an open and rules based global trading system, political support to the General Agreement on Tariffs and Trade (GATT) or World Trade Organization (WTO) round of the day and a shared determination to resist protectionism have been staples of G7 and G8 summit language over decades. As early as 1985, the G7 had also acknowledged the interdependence of growth within the G7 countries and opportunities in the developing world (The Bonn Economic Declaration 1985). Together with the growing need for the G7 and G8 to show responsibility for global issues, this had led to a substantial programme of development commitments with a strong focus on global health (G7 Elmau Progress Report 2015: 88–93). The implementation of social standards came to the fore in Heiligendamm, where G8 leaders stressed that "open markets rest on political acceptance", nurtured by the "social dimension of globalization" (G8 2007 Summit Declaration 2007) Strengthening corporate social responsibility (CSR), including through improved transparency of corporate performance on this count, is among the four action areas here. G8 dialogue with five emerging economies[6] led to a joint understanding that CSR builds confidence in open investment policies and that rules like the OECD Guidelines for Multinational Enterprises "could be fostered and promoted" (Post-Heiligendamm Process Concluding Report 2009). In more general terms, G20 leaders also acknowledged their responsibility for sustainable production and consumption (Annex G20 Leaders Pittsburgh Declaration 2009).

The idea behind the Elmau language on sustainable global value chains was that, against this background, the three elements of the case for action—using individual agency for better standards implementation, building political support for open trade and creating added opportunities for sustainable development—can now fall into place and make a pattern. This idea relies on the fact that a growing number of consumers in G7 countries are prepared to pay a premium for goods produced in a sustainable way (The Nielsen Company 2015: 8). But they cannot translate this preference into actual buying behavior as long as they have no easy way to tell which product on offer is really made sustainably. Product information can be unreliable, confusing due to different overlapping standards, or absent entirely. Due to these

[4]This document also provides a preliminary snapshot on implementation measures taken since the Elmau summit.

[5]Previous G8 positions are still binding on the G7.

[6]Brazil, China, Mexico, India, South Africa.

market imperfections, consumer preferences cannot become price signals and the sustainability premium cannot reach producers, who then do not have added incentives to raise their game. This effect is part of the reason why social and environmental standards are sometimes perceived (and resisted) as mere cost drivers and market barriers (Blowfield and Frynas 2005: 499–513). This could be true if standards would only be rules. But they are also demand factors waiting to catch on. The concept was to let consumer preferences effectively translate into producer incentives. And the joint positions[7] taken in the Elmau Leaders' Declaration span this exact distance: The G7 agree on product level transparency for consumers, better coordination in the way businesses (particularly SME) implement standards and communicate their compliance as well as capacity building in developing producer countries, so that they can actively draw the sustainability premium. In addition, the G7 committed to stronger recourse and remedy mechanisms, so that people affected by infringements of the OECD Guidelines are empowered to become additional drivers for better compliance. Collective G7 action on all those counts was needed to ensure more market clout and less diversion.

The fact that this part of the leaders' language is closely linked to the trade section has already been noted. It is fair to construe this as expanding on the earlier message conveyed at Heiligendamm: The G7 do look at the quality of growth when promoting the trade agenda. And this quality comprises the social and ecological dimensions of sustainability—including in developing countries. That interpretation is corroborated by leaders' support in the same document for what was to become the 2030 Agenda (G7 Elmau Summit Leaders' Declaration 2015: 12). In June 2015, it was already conceivable that sustainable consumption and production could be part of the package.

Will the Elmau commitments make a difference to how business in global value chains is conducted? The effect will not be direct. The enhanced debt relief measures agreed on among the G8 in Cologne have by themselves changed the public finance parameters of highly indebted poor countries (G8 Muskoka Accountability Report 2010: 19). In contrast, the Elmau commitments are designed to enable business to act on their own interest in sustainable global value chains. The effect works through voluntary behavior change. It may be no less powerful for that. First, the implementation of development related G7 commitments is followed up in accountability reports published before every summit. Given the keen interest different communities within global civil society, from human rights and labor to development, take in sustainable global value chains, doing nothing is not a viable option for any G7 country. Second, the Elmau commitments are about better coordination across markets and countries, so that potential first movers do not stop at the prospect of being set back against their competitors. Indeed, the willingness of companies to take part in national multi-stakeholder initiatives has arguably increased with the

[7]Note that not all of them are commitments in the sense that their implementation is formally reported on in future G7 accountability reports. Not every part of leaders' language calls for structured reporting under the formal G7 accountability process.

growth of similar initiatives in other G7 countries and beyond. Finally, Elmau sent a signal that G7 Heads of State and Government are willing to act on sustainable value chains. This signal sets a goal that might be pursued through regulatory action if voluntary business initiative is not deemed effective. So overall there is reason for cautious optimism. Implementation of the Elmau commitments has already begun (G7 Ise Shima Progress Report 2016: 32–34). And business has much to gain from more sustainable global value chains.

2.2 Why Was It Hard

These outcomes were achieved because the G7 agreed on them. While this is a consensus among leaders, it was prepared through a series of intense and productive deliberations on Sherpa[8] and working levels. Discussions improved everyone's understanding and over time built consensus. Crucially, the Elmau outcomes do not ratify a concept prefabricated in Germany. They are a product of genuine co-creation. Getting there was not easy. Four hurdles had to be overcome:

First, the case for action made above does have some inner tensions. Understanding and solving them became an important part of the G7 conversation. They occur on every side of the standards-development-trade triangle:

- *Standards and development:* Standards help to turn trade opportunities into development gains. But they can also limit, divert or stop trade, so that for some, these opportunities do not arise. Many say social and environmental standards are really just a new guise for the old protectionism (Blowfield and Frynas 2005). And there is an inevitable element of self-interest to every serious trade policy. Managing trade-offs between employment goals in the G7 and in third world countries is of course part of that general consideration. Given the political economy of G7 countries it is also easy to imagine that the domestic feel-good conveyed by standards may perhaps override the export concerns of a poor, far-away country. And putting self-interest into normative terms is a time tested bad practice, typical in history for relationships between countries at different stages of development (Schama 1987). This is why the G7 discussed how to turn standards implementation into an honest value proposition for developing producer countries. The Elmau commitment on capacity building and the related ongoing work done by the World Bank are just two outcomes of this conversation.
- *Standards and trade:* Flatly using standards to justify trade openness would have been no less risky. G7 action alone will obviously not change the overall nature of trade in a way that could convince every skeptic. The extent to which trade eventually serves sustainable consumption and production also turns on other

[8]Sherpas are a group of high level civil servants, each preparing his or her leader for G7 summits.

factors: technological change, geopolitics, the scope and contents of future trade agreements beyond the G7 membership. It so became apparent that the G7 could use their commitment to social and environmental standards as a political signal about what kind of trade they want to facilitate. Alternative messages—say a ban on broader types of trade or a promise to make some of the more urgent trade-offs brought about by globalization go away—would have been neither feasible nor credible.

– *Trade and development:* The "necessary but not sufficient" relationship between trade and development has already been noted above. In previous commitments, the G8 had acknowledged trade as a generally positive force for development (G7 Ise Shima Progress Report 2016: 28). This raises the question of what responsibility the G7 take for cases where trade may hurt development. For now, it seems that trade is deemed useful, but not answerable to development. Seen from the trade side, it is the same picture in reverse: Development is obviously useful to trade, but not answerable to it.[9] Apparently, trade and development communities still have different, and at times conflicting, priorities. As a consequence, the G7 designed their commitment on sustainable global value chains to be an effective political impulse using a new leverage point, rather than a vacuous guarantee for more and better trade.

Second, each G7 member first looked at the issue through the lens of their own existing domestic measures for sustainable global value chains. Those are many and various. Approaches range from consumer protection to mandatory corporate reporting to voluntary coordination among businesses to capacity building and policy dialogue in developing countries. And G7 national policies are indeed widely distributed across that spectrum. This is quite natural. What the state is supposed to do (or not do) in imperfect markets for goods and information is a question bound to be answered differently, according to various national traditions of law and economic governance. If a government facilitates the creation of a tool enabling consumers to act on their own value judgments, is it taking sides, effectively backing some producers at the expense of equally legitimate others? Or is it just leveling the playing field, fixing information asymmetries that stifle fair competition? Is it possible for governments to convey a credible message in favor of sustainable consumption and production and still be market neutral? Is there a case for regulation? There can be no general answer to those questions because they turn on the values each country uses to define basic concepts such as "fair" market or "workable competition". In spite of their high levels of integration, remaining structural differences across the G7 testify to that fact. This is why harmonization was out of the question and leaders' language leaves space for different ways to implement the Elmau commitments.

Third, the issue required a mode of operation the G7 was not used to. The Lough Erne summit in 2013 first showed on a substantial scale that G8/G7 development

[9]Whether this approach is still consistent with the vision underlying the Agenda 2030 is a question well worth discussing.

commitments could and indeed should operate through changes in policy rather than earmarked money pledges ("policies not purses"). This approach is now tested but still far from being general practice in the G7. Over decades, G7 development commitments have been prepared by development administrations only. This new agenda touched on different parts of government, from labor to consumer protection to trade to development. In order to engage, every G7 member first had to determine who was in charge and how they would coordinate internally.[10] But also for this cross-government approach, Lough Erne had set a useful precedent. That, along with the ongoing negotiations in New York on new development goals meant to be as relevant at home as abroad, helped to build consensus on an unaccustomed way to take action.

Finally, there were the challenges probably inherent in every G7 process. The German presidency led this conversation, but also carefully accommodated its partner's interests. They were confronted with a range of German goals, different both in their priority for Germany and their appeal to various G7 members. Every partner had to make his own call on where to spend his negotiating capital. And while the sustainable value chain agenda did not entail financial pledges, implementing it would still require some effort. "Policies not purses" is everything but a free ride. This was compounded by the dynamism of international policy making. As new issues continuously came to the fore, no conceivable deal could offer protection from tomorrow's pressing demands. And everyone knew that G7 accountability would make failure to implement commitments just as visible as compliance. Accepting this exposure is a strategic decision. But it also had something of a leap of faith.

2.3 *How It Worked*

It could have failed. While many factors contributed, there was also an element of chance. In hindsight, it is indeed hard to tell pluck from luck and separate the effects of intentional operation from those of propitious coincidence. With this caveat, four groups of causes must be credited:

- First, structure mattered. The G7 is a small group of countries similar to each other in many ways. They are bound together through a dense network of working relationships in different fora and on various levels. This allows for easy and frequent exchanges, even in the absence of a formal working group in charge. Because there were also no regular ministers meeting on the matter before the summit, the issue had direct access to Sherpas from the start. Close cooperation between development and labor ministries in Germany was critical and

[10] Questions like "Is this a labor or a development issue?" were raised but not easily answered.

helped to broaden support. Labor, development and human rights communities across the G7 were mobilized.

- Second, the time was right. People in G7 countries and beyond continue to question the way global trade works. To many, the collapse of Rana Plaza still epitomizes a system where business profits from value created abroad but leaves social and environmental costs to others. Again, ongoing negotiations on the SDG focused minds on how every nation, regardless of its income level, really takes part in one global development process. Finally, events in Greece and Ukraine generally raised expectations on Germany's ability to facilitate solutions.
- Third, for all of its tensions, the case for action was and still is strong. Germany argued that business success and standards implementation can and often do go together. The issue was what the G7 can do to help scale these cases. Among other effects, this debate led to a promising strain of ongoing analytic and advisory work at the World Bank. Germany also gave reasons for collective action. Some G7 members had already started relevant national measures. But first movers are set back against competitors without some coordination.[11] The G7 was a place to start. Even the new type of commitment ("policies not purses") helped in a way. Policy implementation can carry its difficulties, but at least this conversation was free from the intricacies of cost assessments and burden shares. On a practical level, valid questions were left open if they did not affect what the commitment was meant to achieve. Open wording was important to make sure implementation could work through different national approaches, but still make a coherent whole.
- Fourth, there was work. Political leadership in Germany was from the start strongly committed to achieve G7 consensus on this issue. G7 partners understood that Germany would engage them on this throughout its presidency. Gradually, as the case was made and the possibility of success became visible, more decision makers came to share that commitment. The core of the work lay in the many conversations with G7 partners, bilaterally and in group format. This process helped to understand the nature of the interest each partner takes in the agenda, the scope and limitations of what they can contribute, their specific needs concerning the outcome and the origins of their doubt. This series of exchanges tested the concept, putting a spotlight on its weaknesses but also showing where it had gained traction. In this sense, meeting minutes added up to a hiking map, showing dead ends, creeks and bridges on the way to Elmau. Starting early and keeping your direction was important. Early engagement with G7 partners and frequent continuous follow-up on a consistent agenda paid off in the end. This approach gave partners time to get familiar with the concept. Repeated opportunities to position themselves and the fact that Germany did not move the

[11] German business had made a convincing case that government action on sustainable global value chains should by nature be coordinated with international partners.

goalposts made every engagement build on the previous one and let everyone develop ownership of the outcome. Finally, two high profile events served to create visibility for the agenda and mobilize further support: A stakeholder conference in March 2015, hosted by Germany's development and labor ministers, attended by the heads of the World Bank, the International Labor Organization (ILO) and the OECD and featuring a keynote speech by Peace Nobel Prize Winner Kailash Satyarthi. And the joint meeting of G7 labor and development ministers already mentioned above.

3 Some Open Questions—and a Piecemeal Conclusion

The concept underlying the Elmau commitments for sustainable global value chains is designed to operate within the current global economic environment. But various trends continue to change that environment and again affect every side of the trade-development-standards triangle. This final part will attempt to picture some of these trends and raise questions about their potential effect on the goals set in Elmau.

3.1 The Impact of Digitalization

One business reason to integrate developing country producers into global value chains is their lower coordination cost. It allows for larger firm scopes in developing countries, comprising different business functions which would be unbundled in a lower transaction cost environment. Perhaps many of the development gains projected through global value chains (such as skills development) are mediated through the economic and social integration processes that happen when firms grow in scope and size and take on many functions. The moment coordination costs can be lowered without cross-border production and without building large firms, this effect does not work anymore.

Digitalization may be that moment. The control many firms exert on global value chains is already informational in nature. To the extent that programmed automation displaces human agency in business organization or production, the entire process of creating value for markets can be concentrated wherever the information lies ("digital reshoring"; Cowen 2016). The Adidas Speedfactory (BMWi 2014: 36f) may be a glimpse of more to come. Even if reshoring may be more or less relevant depending on sectors, it is clear that at least some of today's trade-development links will work differently in the future. In some cases, present challenges of implementing social and environmental standards could then just fall away, because countries with weak legal infrastructures would not even be part of the process. Obviously, this is not the outcome standards proponents had in mind. It shows how the case for standards implementation and the case for development really belong together. Seen in isolation, the very nature of the standards case, normative and "yes or no" as it is, tends to render answers that are just meaningless in development, "one after the other" terms: Either you comply and then probably do not need to develop. Or you do not and your noncompliance serves as a reason not to do anything about it.[12]

The development opportunities inherent in digitized production are as yet far from being understood. But at least two aspects emerge: First, development in many middle income countries has been so fast that they might become lead locations for digitized production. Alibaba founder Jack Ma feels that Asian original equipment makers already offer better product than the global brands they cater to (Financial Times, June 14th, 2016, see also Taglioni and Winkler 2016: 14). Second, expanding connectivity and network effects might generate more access to digital resources and services in ever larger parts of the developing world. This last effect may possibly also help the transmission of price signals from conscious consumers to responsible producers.

3.2 Changes, Cyclical and Structural

Apart from digitalization, global value chains will have to adjust to other forces: Low global demand and technological change may lead to continued cost pressure. Geopolitical tensions, the rise of populist nationalism, concerns of access to resources and the need for protection from the effects of climate change may all motivate firms to limit their exposure and make value chains less international.

The Elmau concept works because there is a critical mass of consumers willing to pay for sustainability (The Nielsen Company 2015: 8) and because, through transparency and coordination, there will be better channels to transmit that price signal to producers. Its global impact rides on the number and buying power of conscious consumers as well as on businesses' readiness to take that opportunity. Both conditions point to the growing consumer and factor markets in middle income countries. Most social and environmental standards are notionally universal. However, they largely originate in the European welfare state and it shows. This is why reaching an ambitious consensus on standards implementation beyond the G7 is not easy. But it is possible. While the functioning of worker health insurance is inevitably contingent on a country's legal system, the aspiration behind it—to collectively protect people from hazards they cannot control—is truly universal. The development, "one after the other" approach to standards will be crucial to make that point. Just saying we were there first will not produce a useful conversation. A more promising way could be to identify and measure the positive impacts a given business activity within a global value chain has on various development goals. Whether and at what speed the rising middle classes in emerging economies will take

[12]Development economics discusses similar conundrums—the difficulty of telling drivers from outcomes of development—under the notion of endogeneity.

more interest in sustainable consumption is hard to tell. On the one hand, newcomers to the middle class will understandably want to stabilize their position before thinking about social and environmental externalities. On the other hand, those very externalities exert growing pressure on the quality of life, particularly in middle income countries (The Nielsen Company 2015: 07). This triggers demand for government action. But emerging societies also seem to have a keen sense of what individual agency can do. Finally, the way inequality in income and wealth develops in emerging countries will impact on where responsibility for larger social and environmental issues is seen. Whoever feels he has no control over his own life will not readily take responsibility for others.

Branko Milanovic (Milanovic 2015: 452–460) has shown how much of the difference between global incomes is associated with nationality. His term for this observation is "citizenship rent". His idea is that it pays to belong to a rich country and that today you profit from that way more than you could gain from just being productive. In this framework resistance in rich countries against new trade agreements may readily be construed as an attempt to defend the citizenship rent. If this were true, a commitment to sustainable consumption and production on a global scale would by itself not be able to overcome skepticism about trade. And better implemented social and environmental standards, in this view being part of the citizenship rent, would then certainly be used to block rather than improve trade.

This is why the agenda for sustainable global value chains should find ways to demonstrate material gains not only for developing producer countries but for workers in rich countries as well. Among other factors, this will depend on how gains from more efficient sourcing are distributed within the firm and the societies where it operates. Although vertical competition for production tasks will not disappear, it is worth exploring where international division of labor can take new forms that every partner can use for his development. The agenda is far from being exhausted.

3.3 A Piecemeal Conclusion

Where does all this leave us? Trade growth is far from continuous (IMF 2016: 63) and participation in global value chains seems to have leveled off (IMF 2016: 82). But for all the drivers of change set out above, global value chains will not go away soon. Digital production at home does not equal autarchy. Every economy will still need to access things that physically exist abroad (resources, commodities, specialized labor, demand) and then transform them using its own respective abilities. In this sense, global value chains will probably continue offering positive development opportunities for poor countries. Tapping consumer demand for sustainably made goods and using the improvements and efficiencies brought about by compliance are likely to remain among those opportunities.

However, finding, expanding and protecting one's own niche in global value chains and using that niche for sustainable development may require different strategies in the future. First, countries may need to limit their exposure to digital reshoring. Comparative advantage in trade may then increasingly be tied to the physical, immoveable assets of an economy, because they cannot easily be displaced or substituted through digital means. As it happens, sustainability in production eventually succeeds (or fails) on this very level—the cultural and physical endowment upon which every economy operates.

Second, countries could explore where they can position themselves not only as producers and sellers, but increasingly also as buyers and users along digitally enhanced global value chains. While producer and supplier roles will remain crucial steppingstones for development, low-cost or no-cost access afforded by innovative digital services can be used as an additional strategic asset.

Third, precisely because they have spread across continents and radically cut timespans[13] within and between every step of the production process, global value chains have become vulnerable to shocks from an increasing number of sources. For a risk assessment, see: Chartered Institute of Procurement and Supply (CIPS), CIPS Risk Index Quarterly Report Q3 2016, Stamford/UK 2016 (CIPS 2016). This is why investments in the resilience of global value chains may now yield a higher marginal return than chasing the next efficiency. Just as every stakeholder, developing producer countries have a role, as well as an interest, in building more robust global value chains. Depending on a country's particular position, this role can have different scripts: Building up domestic markets and perhaps positioning yourself as an important buyer at a strategic point within a given global value chain. Deepening regional integration with neighbors is less likely to disrupt trade flows. Diversifying into different niches, so that expanding one can offset the eclipse of another and carefully managing trade imbalances are also important. Sustainable business practices themselves may make global value chains more robust. They help prevent social and environmental disruption and convey a consistent value proposition to consumer markets.

References

Annex G20 Leaders Pittsburgh Declaration (2009) Core values for sustainable economic activity. http://www.g20.utoronto.ca/2009/2009communique0925.html#annex. Accessed 20 June 2017

Blowfield M, Frynas JG (2005) Setting new agendas: critical perspectives on corporate social responsibility in the developing world. Int Aff 81(3):499–513

Bundesministerium für Wirtschaft und Energie (BMWi) (2014) Autonomik für Industrie 4.0, Berlin 2014. https://www.bmwi.de/Redaktion/DE/Publikationen/Digitale-Welt/autonomik-fuer-industrie-4-0.pdf?__blob=publicationFile&v=9. Accessed 20 June 2017

Chalmers D et al (2010) European union law, 2nd edn. Cambridge University Press, Cambridge

[13]The "just in time" (JIT) paradigm in production and logistics goes back to a process pioneered by Toyota. It will probably further gain in relevance through the "internet of things" (IOT).

Chartered Institute of Procurement and Supply (CIPS) (2016) CIPS risk index quarterly report Q3 2016. https://www.cips.org/Documents/cips-for-business/CIPS_D_B_Risk_Index_Report_Q3_13pp_A4_1016_v3.pdf. Accessed 20 June 2017
Cowen T (2016) Economic development in an average is over world, 8 April 2016
Fally T, Hillberry R (2015) A coasian model of international production chains, World Bank policy research working paper no. 7435, Washington, DC
G7 Elmau Progress Report (2015) G7 Elmau progress report
G7 Elmau Summit Leaders' Declaration (2015) Leaders declaration G7 summit 7–8 June 2015. https://sustainabledevelopment.un.org/content/documents/7320LEADERS%20STATEMENT_FINAL_CLEAN.pdf. Accessed 20 June 2017
G7 IseShima Progress Report (2016) Ise-Shima
G8 2007 Summit Declaration (2007) Growth and responsibility in the world economy, G8 summit declaration (7 June 2007). http://www.g8.utoronto.ca/summit/2007heiligendamm/g8-2007-economy.pdf. Accessed 20 June 2017
G8 Muskoka Accountability Report (2010) G8 Muskoka accountability report – assessing action and results against development-related commitments. http://www.mofa.go.jp/policy/economy/summit/2010/pdfs/accountability.pdf. Accessed 20 June 2017
Hay P (2002) Law of the United States. C.H. Beck, Munich
International Monetary Fund (IMF) (2016) World economic outlook. http://www.imf.org/external/pubs/ft/weo/2016/02/. Accessed 20 June 2017
Milanovic B (2015) Global inequality of opportunity: how much of our income is determined by where we live? Rev Econ Stat 97(2):452–460
Post-Heiligendamm Process Concluding Report (2009) http://www.g8.utoronto.ca/summit/2009laquila/2009-g5-g8-1-hdp.pdf. Accessed 15 June 2017
Schama S (1987) The embarrassment of riches: an interpretation of Dutch culture in the Golden Age. Knopf, New York
Taglioni D, Winkler D (2016) Making global value chains work for development. World Bank Group, Washington, DC
Terazono E (2015) Sustainability in commodities accrues momentum. Financial Times
The Bonn Economic Declaration (1985) Towards sustained growth and higher employment. http://www.mofa.go.jp/policy/economy/summit/2000/past_summit/11/e11_a.html. Accessed 20 June 2017
The Nielsen Company (2015) The Nielsen global survey of corporate social responsibility and sustainability 2015. The sustainability imperative, London
Transforming Our World (2015) The 2030 agenda for sustainable development (SDG), New York

Transforming the Garment Industry in Bangladesh: Sharing Responsibility

Dorothée Baumann-Pauly, Sarah Labowitz, and Nate Stein

1 Introduction: Chronic Human Rights Issues in Bangladesh's Ready-Made Garment Sector

Working conditions in Bangladesh's Ready-Made Garment (RMG) sector have been known to be precarious for decades. Throughout the history of the sector's rapid growth, international organizations, multi-stakeholder initiatives and individual brands have tried to improve labor rights and health and safety standards in the industry, yet with limited success.

After a series of accidents in previous years (CBC News 2013), the collapse of the Rana Plaza factory complex in the outskirts of Dhaka in April 2013 represented the deadliest incident in the history of the garment industry. Over 1100 workers lost their lives and over 2500 were injured.

The international attention that followed Rana Plaza created momentum for the formation of two unprecedented collective initiatives. The Bangladesh Accord on Fire and Building Safety (2017), and the Alliance for Bangladesh Worker Safety (2017). Both initiatives united brands and retailers around a set of standards for fire and building safety and created monitoring mechanisms for tracking progress.

These initiatives for the first time brought together major brands and retailers to jointly address long-standing governance gaps around health and safety issues in garment production sites. Such an industry-wide collaboration is remarkable and many social science academics chose them as their subject of analysis due to the laboratory-like conditions in Bangladesh after Rana Plaza. A stream of research is currently underway and the overall tone of existing academic analysis is positive (Anner 2015; Lund-Thomsen and Lindgreen 2013; Reinecke and Donaghey 2015).

Our in-depth research on the business models and incentive structures in this industry, however, revealed that the supply chain of garments in Bangladesh and its

D. Baumann-Pauly (✉) · S. Labowitz · N. Stein
NYU Stern Center for Business and Human Rights, New York, NY, USA

S. M. Jastram, A.-M. Schneider (eds.), *Sustainable Fashion*, Management for Professionals, https://doi.org/10.1007/978-3-319-74367-7_4

human rights challenges are much more complex than the two private initiatives currently acknowledge. From two separate research reports in 2014 and 2015, we conclude that on their own, these mechanisms are insufficient to protect all workers in Bangladesh's garment sector (Labowitz and Baumann-Pauly 2014, 2015).

In this chapter we outline the remaining challenges of the garment sector in Bangladesh. We argue that transformative change in industries with seemingly intractable human rights challenges requires that all stakeholders commit to share responsibilities. No actor alone can fix an industry with as many challenges as the garment sector in Bangladesh. We refer to a shared responsibility concept for global supply chains that has been developed in the context of the World Economic Forum's Global Agenda Council on Human Rights. In so doing, we use Bangladesh and this specific sector as an emblematic case to answer the broader question in this context: How can we develop an approach for transformative change in industries with chronic human rights issues?

1.1 Shortcomings of Existing Initiatives

Limited scale—current business models of brands and retailers set incentives for opaque subcontracting networks.

Our research on the garment sector has shed light on the business models that brands and retailers employ to source from Bangladesh (Labowitz and Baumann-Pauly 2014, 2015). We were able to show that given intense pressure on price and production times, subcontracting in the industry is an efficient and rational business practice. As a result, we estimate that the industry is much larger than currently assumed, with many smaller production facilities supporting suppliers that work directly with brands.

Subcontracting is by no means a modern innovation. At the turn of the twentieth century in New York's garment district, almost all clothing production happened in workshops which contracted with larger "manufacturers." The label on a shirt or a dress simply indicated that the pieces had been cut from the company's pattern and that the finished garment had been inspected before sale (Von Drehle 2003). Our research in Bangladesh confirms the persistence of this system on a global scale.

Subcontracting is a high risk factor for sweatshop conditions. With every additional layer between the workers and the final buyer, margins become thinner. In the production of apparel, the only flexible cost factor is labor cost and subcontracting squeezes this cost factor (Labowitz and Baumann-Pauly 2015). Subcontracting is typically contractually forbidden unless authorized by the brand. Suppliers risk major fines or terminations of the business relationship with the brand if they subcontract without authorization. However, current sourcing practices with notoriously low prices and increasingly shorter lead times often force suppliers to cut costs. As a result, subcontracting is largely a hidden practice that creates opaque networks that are not monitored by brands and workers in deeper layers are the least protected. Given that the cost and delivery time pressures are the main reasons for subcontracting, it is also very likely that workers at subcontracting facilities are earning less, work longer hours, and in case of mistreatment, they have no place for redress (Van Heerden 2010).

Despite the prevalence of subcontracting, current remediation efforts in Bangladesh only reach the first layer of the supply chain. While the Accord and the Alliance work closely with about 1900 direct suppliers, our research shows that there are more than 7000 garment factories in Bangladesh (Labowitz and Baumann-Pauly 2015). Following our investigation, local institutions like BRAC University and the industry association Bangladesh Garment Manufacturers and Exporters Association (BGMEA) have recently conducted their own research on the number of RMG factories in Bangladesh (Munni 2016). At a closer look, both studies also indicate high levels of subcontracting (Labowitz 2016).

Limited scope—current initiatives focus narrowly on fire and building safety and do not address systemic issues that undermine compliance with safety standards.

Systemic issues in Bangladesh, for example the need to improve a dangerously poor electrical grid, threaten to undermine the efforts that are underway to address fire and building safety (Labowitz and Baumann-Pauly 2014). Heavy diesel generators that many production sites installed to supplement the notoriously unreliable power supply, strain building stability and contributed to the collapse of Rana Plaza (Manik and Najar 2015). Similarly, short circuits have caused fires at factories (All Jazeera 2012).

Focusing on fire and building safety at production sites alone fails to identify systemic issues that are critical for establishing safe workplaces. Infrastructure upgrades are beyond what brands and retailers can deliver in Bangladesh (or elsewhere), yet such measures are indispensable for real progress in securing safe workplaces. Defining the challenges for fire and building safety comprehensively is necessary to reach the widely shared objective of a safe and sustainable industry.

Moreover, the brands' narrow focus on fire and building safety may cause other chronic labor rights issues to diminish in the background. Issues such as excessive overtime, constraints on freedom of association (Clean Clothes Campaign 2017a, b), discrimination, and non-payment of wages remain unresolved in Bangladesh, as in many other low-wage environments (Business Social Compliance Initiative 2014).

Improving working conditions and meeting international labor standards even in low-wage economies will require an integrated approach that addresses the nexus between hours of work, wages and business models. Some multi-stakeholder initiatives such as the Fair Labor Association, have created work streams to discuss purchasing practices with their participating brands (Fair Labor Association 2012), and other initiatives attempt to create innovative platforms that allow suppliers to rate the purchasing behavior of their clients (Better Buying 2016). Yet at this point, these initiatives are not mainstream and the situation for workers in the industry remains difficult. The lack of attention on the broader basket of labor rights issues is a shortcoming of existing initiatives in Bangladesh.

Limited remediation success—current initiatives do not meet their own objectives.

Aside from the Accord and Alliance's limited scale and scope, the two initiatives have struggled to implement remediation plans in the 2334 factories they have inspected. Their own remediation reports show major delays in implementing corrective action plans (Accord Progress Overview 2017; Alliance Progress 2017). While nearly all Accord and Alliance factories have been inspected, only 79 have passed final inspections as of March 2017 (NYU Stern Center for Business and Human Rights 2017). There are still over 65,000 individual electrical, fire, and structural safety problems left to be addressed in the Accord factories alone (Accord Progress Overview 2017), and progress has been similarly slow through the Alliance (Alliance Progress 2017). New research based on the progress reports of Accord and Alliance has shown that less than 4% of factories had fully completed the initiatives' correction action plans (NYU Stern Center for Business and Human Rights 2017).

With limited time left before the 5-year deadline, it is possible that both brands and their suppliers are seeking to "run out the clock" on the remediation initiatives. According to an ILO/IFC study, the estimated cost for remediation per factory ranges between $100,000 and $250,000 U.S. Dollars (International Labour Organization 2016), and in extreme cases up to $1.5 million dollars (Mathews 2016). The failure to address the question of who pays for remediation is considered the biggest barrier for implementation (Russell 2016).

Due to the fact that the factory remediations are lagging far behind public expectations with only a year left in the five-year plan, both initiatives have started thinking about extending their mandate beyond 2018. They could do that without the consent of local authorities but this would certainly not help the implementation. Alternatively, the initiatives could try to negotiate with the industry associations and the Bangladesh government and offer to include them in future activities. Neither government representatives nor industry associations are at this point part of the decision-making bodies of those two initiatives (The Daily Star 2016). Neither option, however, proposes an institutional design that could assess and address the issues in Bangladesh's garment sector comprehensively.

Despite these challenges, garment exports from Bangladesh have been growing between 2013 and 2016 (Exports of Goods and Services Annual Percentage Growth 2017), supported by demand as well as international trade incentives such as the EU's Everything But Arms (Rutledge 2016), and more workers than ever are involved in production (Trade Information 2011). How can the safety and dignity of these workers be protected effectively?

2 Elements of Transformative Change

Transforming Bangladesh's garment sector requires concerted efforts. No stakeholder group alone can effectively address the full scale and scope of Bangladesh's garment business. To outline an approach for developing a transformative strategy, we apply the elements of the shared responsibility model to the context of Bangladesh's garment sector (Posner 2015). The model essentially rests upon three elements—an assessment of the full scale and scope of the problem, an assessment of the cost of fixing the problem, and a commitment by key stakeholders to a process for defining how to share responsibilities and costs.

2.1 Comprehensive Assessment of Outstanding Challenges

Transformative change that renders the sector safe and sustainable will require a comprehensive problem assessment. Most importantly, all facilities involved in the production of garments in Bangladesh need to be identified, mapped, registered and assessed. Given the fluid relationships between suppliers that have direct relationships with brands and retailers (Labowitz and Baumann-Pauly 2015), and facilities that subcontract from direct suppliers, it is not useful to strictly differentiate between those two types in the mapping and assessment process. For a safe and sustainable industry, all garment facilities need to be able to meet minimum standards.

Visibility over the entire scope of the industry, including human rights challenges, is the goal, yet that visibility should not be confused with responsibility (Posner 2015). Many brands fear that striving for end-to-end transparency in their supply chains will make them automatically responsible for fixing the issues that this process will reveal. They therefore focus on first-tier suppliers where they have the greatest leverage to change conditions, despite possibly more urgent human rights risks in deeper layers of their supply chain. Decoupling visibility and responsibility can therefore provide the safe space that is necessary for collecting the information for a complete mapping of human rights risks in the supply chain. Civil society, unions, the media, and academia have an important role to play in the process of gaining visibility. They can help to collect the information necessary for a complete mapping.

Our centre at New York University (NYU) has piloted a mapping project of the garment sector in Bangladesh (NYU Stern Center for Business and Human Rights 2015) to demonstrate that greater transparency in the sector is possible and necessary to adequately assess the sector. Without understanding the full scale and scope of the issues (including the root causes of systemic human rights issues) devising transformative change strategies will neither be feasible nor successful.

2.1.1 Assessing the Total Cost of Remediation

As explained above, achieving more comprehensive visibility does not mean that companies alone must bear the costs of resolving every risk identified. Rather, greater visibility will give companies, governments and other stakeholders the information they need to more effectively identify and prioritize the areas that need the greatest attention and to craft collective solutions (Posner 2015).

Once all garment facilities are mapped and their remediation needs are assessed, the cost of fixing the issues can be estimated. The estimation requires an honest assessment of the full scope of the industry, the necessary technical improvements, and the steps necessary to ensure respect for labor rights. Estimating the total cost is necessary for fully understanding the funding needs and for developing funding strategies.

Deciding who will eventually be responsible for fixing issues and who will bear the cost requires dialogue processes with all stakeholders in this value chain to establish the fair share for each (ibid.). Thus, while responsibilities are shared among stakeholders, they are also clearly differentiated; the special responsibilities of each stakeholder must be specified and agreed upon in order to pursue the group's collective goal (ibid.). A multi-stakeholder taskforce of experts under local leadership may be best suited to provide a framework for a discussion over how to distribute responsibility and costs.

The Blight Removal Task Force in Detroit provides one example of what can be achieved when governments, and private ventures work together to address a problem, and how funding can be raised collaboratively. In Detroit, Quicken Loans, a private mortgage lender, has partnered with state and local governments to address residential blight. Together they developed a comprehensive plan of how to map and remove blighted buildings (Detroit Blight Removal Task Force plan 2014). This partnership has raised the funds for fixing the problem—nearly $1 billion with the goal of spurring Detroit's economy. The lessons from Detroit could be applied to Bangladesh's garment sector that also needs a holistic approach to ensure sustainable solutions.

Beyond funding, collaborative efforts among stakeholders also help the various parties pool their knowledge, resources, and leverage to find practical remedies and feasible solutions (Posner 2015).

2.1.2 Collaboration with a Greater Pool of Stakeholders and Local Actors with the Bangladesh Government in the Lead

The magnitude of the challenges in Bangladesh's garment sector requires widening the range of actors that are currently involved in driving change. Local actors—factory owners, unions, and political leaders—are especially important constituencies that have not been fully integrated in existing initiatives. Local elites and local manufacturers have a key role to play to increase supply chain transparency and assume responsibility for their own back-supply chain (Huq 2014).

International financial institutions also need to play a more active role in facilitating practical funding solutions for upgrading the sector. The World Bank, the International Finance Corporation (IFC), and other international financial institutions could, for example, help underwrite infrastructure improvements in key sectors. They could also help to set up the infrastructure for shared financial responsibilities, for example through a common fund that will be used to address the human rights problems or risk in question. They could also facilitate access to funding, which for instance has been an issue for the implementation of the corrective action plans of the Accord and the Alliance (Russell 2016).

Local government actors also need to contribute to sustainable improvements by strengthening the regulatory framework in which the apparel industry is embedded. The Rana Plaza tragedy did speed up the adoption of a new Bangladesh Labor Act on July 15, 2013, but its full implementation requires time. For example, since the adoption of the new Labor Act, labor unions are technically able to form freely in Bangladesh outside of export processing zones, yet the labor union movement to date is nascent and requires further support to truly represent workers' interests. The International Labor Organization (ILO), certain brands, and several projects funded by donor countries have focused on strengthening emerging union organizations and workers' voices at individual production facilities. Yet local actors are still not leading these efforts, a factor that may hamper their sustainability. Similarly, in the organizational structure of the Accord and the Alliance, key actors from Bangladesh are sidelined or not part of the decision-making body at all. Local manufacturers, for example, the actors that are supposed to implement the remediation plans, are not included in the Accord's board (Alliance Board of Directors 2017; Steering Committee Members 2017), and resistance to the implementation has been an issue.

3 The Apparel Sector as an Emblematic Case Study for Addressing Chronic Human Rights Issues in Global Supply Chains

The most serious human rights risks in certain supply chains often are common to particular products or commodities and specific to certain geographical areas where governance or sourcing practices are challenging. As a result, industry-specific, rather than company-specific, approaches are critical for addressing the most serious human rights risks and challenges (Posner 2015). In many emerging economies, the apparel sector is the most important industrial sector, for example, clothing exports make up 82% of all exports in Bangladesh, 62% of all exports in Lesotho, and 39% of all exports in Malaysia (Exports of Goods and Services Annual Percentage Growth 2017). Developing nations generally move away from an economy based on agriculture into one based on low-skill manufacturing, before attempting (but not always succeeding) another transition to high-skilled manufacturing and services. Additionally, wages are usually higher in manufacturing than in agriculture.

Apparel production has also been under attack for decades by critical civil society organizations for being dangerous and exploitative (The Clean Clothes Campaign 2017a, b; Labour Behind the Label 2017; Moore 2015). Despite a steep learning curve for some brands (Zadek 2004) and the tremendous efforts of many stakeholders to improve conditions, the human rights situation in the sector remains dire for many workers (Poulton et al. 2014).

This situation is not unique for the global apparel supply chain. There are many other sectors that suffer from chronic human rights issues. Child labor in cocoa production in the Ivory Coast, forced labor in cotton production in Uzbekistan, and slave labor in the Thai fishing industry are all well-known human rights challenges that, despite the attention that they have received in recent years, cannot point to major breakthroughs.

The context of these human rights issues in the supply chain is strikingly similar: national governments are unable or unwilling to provide basic protection; brands sourcing from such countries claim to be unable or are unwilling to look closely into human rights issues. How to share responsibility for establishing decent workplaces in a global economy is therefore a key question that transcends this sector-specific analysis in Bangladesh.

Multi-stakeholder Initiatives (e.g. the Fair Labor Association) that work in these sectors have contributed to some progress, yet they fall short when it comes to addressing systemic issues. Improvements through approaches that focus on establishing compliance at individual production facilities are by definition insular. There is also a growing consensus over the limitations of what can be achieved through social audits (van Heerden and Baumann 2005). Longitudinal empirical data shows that improvements on many complex issues like overtime have been limited and even stagnating. Sensitive issues, in contrast, like harassment and abuse, may not even surface in social audits. The shared responsibility model offers a bold alternative to the still dominant audit methodology with the objective of upgrading an entire sector.

The next step therefore must be to test and refine the shared responsibility approach in practice. There are already some practical initiatives that feature elements of the shared responsibility approach, including the Detroit Blight Removal Task Force, mentioned earlier, as well as the architecture of the Sialkot Soccer Project, which shows how a broad coalition of state and non-state actors can collaborate to address one specific issue, in this case child labor. Another model of shared financial responsibility includes the June 2015 Vision Zero Fund, which uses contributions from governments and businesses to apply to improvements in human rights in supply chains (International Labour Organization 2017).

Defining an approach for sustainable supply chains in the apparel sector can provide insights for how to address other supply chain related chronic human rights issues in other sectors. Our in-depth analysis of the sector in Bangladesh therefore serves as an emblematic case study, not only for helping to define what is termed "sustainable fashion" but possibly also for "sustainable chocolate" or "sustainable fishing" and other industries that have been struggling to operate in ways respectful of human rights. In the human rights context, a volume on sustainable fashion can, therefore, point far beyond its own industry.

References

Accord Progress Overview (2017) http://bangladeshaccord.org/progress/. Accessed 27 Apr 2017

Alliance Board of Directors (2017) http://www.bangladeshworkersafety.org/who-we-are/leadership/board-of-directors. Accessed 24 Apr 2017

Alliance for Bangladesh Workers Safety (2017) http://www.bangladeshworkersafety.org/. Accessed 24 Apr 2017
Alliance Progress (2017) http://www.bangladeshworkersafety.org/progress-impact/alliance-statistics. Accessed 21 May 2017
Anner M (2015) Worker resistance in global supply chains. Int J Lab Res. http://lser.la.psu.edu/gwr/documents/2015IJLRWorkerResistanceinGSCs.pdf. Accessed 21 May 2017
Bangladesh Accord on Fire and Building Safety (2017) http://bangladeshaccord.org/. Accessed 24 Apr 2017
Better Buying (2016) http://www.betterbuying.org/. Accessed 24 Apr 2017
Business Social Compliance Initiative (2014) Annual report. http://www.bsci-intl.org/sites/default/files/bsci-2014-05-26-annual_report_2013-uk-a4-mai_2014-def-internet.pdf. Accessed 24 Apr 2017
CBC News (2013) Timeline of deadly accidents in the Bangladesh garment sector. http://www.cbc.ca/news2/interactives/timeline-bangladesh/. Accessed 24 Apr 2017
Clean Clothes Campaign (2017a) http://www.cleanclothes.org/. Accessed 24 Apr 2017
Clean Clothes Campaign (2017b) Leading retailers withdraw from Bangladesh garment industry's annual showcase event. https://cleanclothes.org/news/2017/02/22/leading-retailers-withdraw-from-bangladesh-garment-industry2019s-annual-showcase-event. Accessed 21 May 2017
Detroit Blight Removal Task Force Plan (2014) Every neighborhood has a future. . . and it doesn't include blight. http://s3.documentcloud.org/documents/1173946/detroit-blight-removal-task-force-plan-may-2014.pdf. Accessed 21 May 2017
Exports of Goods and Services Annual Percentage Growth (2017) World Bank Data. http://data.worldbank.org/indicator/NE.EXP.GNFS.KD.ZG?locations=BD. Accessed 30 June 2016
Fair Labor Association (2012) Respect survey sheds light on purchasing habits. http://www.fairlabor.org/blog/entry/respect-survey-sheds-light-purchasing-habits. Accessed 24 Apr 2017
Huq R (2014) Bangladesh hunts for 29 cents. http://www.wsj.com/articles/SB10001424052702303825604579518333605948014. Accessed 24 Apr 2017
International Labour Organization (2016) Remediation financing in Bangladesh's ready made garment sector. https://www.ifc.org/wps/wcm/connect/650e70d9-3744-41ff-875a-a790840b23b8/Report_Remediation+Financing_Bangladesh.pdf?MOD=AJPERES. Accessed 21 May 2017
International Labour Organization (2017) Vision zero fund. http://www.ilo.org/safework/projects/WCMS_517539/lang--en/index.htm. Accessed 22 May 2017
Al Jazeera (2012) Fire hit second Bangladesh garment factory. http://www.aljazeera.com/news/asia/2012/11/201211266021814214.html. Accessed 24 Apr 2017
Labour Behind the Label (2017) http://labourbehindthelabel.org/. Accessed 24 Apr 2017
Labowitz S (2016) New data on the number of factories in Bangladesh. Available via New York University Stern Center for Business and Human Rights Blog. http://bhr.stern.nyu.edu/blogs/-data-on-number-of-factories-bd. Accessed 24 Apr 2017
Labowitz S, Baumann-Pauly D (2014) Business as usual is not an option: supply chains and sourcing after Rana Plaza. Available via NYU Stern Center for Business and Human Rights. http://www.stern.nyu.edu/sites/default/files/assets/documents/con_047408.pdf. Accessed 21 May 2017
Labowitz S, Baumann-Pauly D (2015) Beyond the tip of the iceberg: Bangladesh's forgotten apparel workers. Available via NYU Stern Center for Business and Human Rights. http://static1.squarespace.com/static/547df270e4b0ba184dfc490e/t/5672bef1c647ad862c494795/1450360561022/Bangladesh-Report-Final.pdf. Accessed 21 May 2017
Lund-Thomsen P, Lindgreen A (2013) Corporate social responsibility in global value chains: where are we now and where are we going? J Bus Ethics. https://link.springer.com/article/10.1007/s10551-013-1796-x. Accessed 21 May 2017
Manik JA, Najar N (2015) Bangladesh police charge 41 with murder over Rana Plaza Collapse. https://www.nytimes.com/2015/06/02/world/asia/bangladesh-rana-plaza-murder-charges.html. Accessed 24 Apr 2017

Mathews B (2016) One in four Bangladesh factories face major remediation costs. Ecotextile News. https://www.ecotextile.com/2016062322198/materials-production-news/1-in-4-bangladesh-factories-face-major-remediation-costs.html. Accessed 21 May 2017
Moore B (2015) 'The True Cost' documentary tallies global effect of cheap clothes. http://www.latimes.com/entertainment/movies/la-et-mn-true-cost-cheap-clothes-documentary-20150528-story.html. Accessed 24 Apr 2017
Munni M (2016) One-fourth of BGMEA units now non-existent or closed. http://www.thefinancialexpress-bd.com/2016/06/14/34040/One-fourth-of-BGMEA-units-now-non-existent-or-closed. Accessed 24 Apr 2017
NYU Stern Center for Business and Human Rights (2015) Garment factories in Bangladesh. http://people.stern.nyu.edu/twadhwa/bangladesh/maps.html. Accessed 24 Apr 2017
NYU Stern Center for Business and Human Rights (2017) Bangladesh factory safety – four years after Rana Plaza. https://static1.squarespace.com/static/547df270e4b0ba184dfc490e/t/590761ded2b8574b0003acf2/1493656030884/Funding+Timeline+FinalVersion.pdf. Accessed 21 May 2017
Posner M (2015) Shared responsibility: a new paradigm for supply chains. Available via World Economic Forum. http://www3.weforum.org/docs/WEF_GAC_Supply_Chains_%20A_New_Paradigm_2015.pdf. Accessed 24 Apr 2017
Poulton L, Panetta F, Burke J, et al (2014) The shirt on your back. http://www.theguardian.com/world/ng-interactive/2014/apr/bangladesh-shirt-on-your-back. Accessed 24 Apr 2017
Reinecke J, Donaghey J (2015) After Rana Plaza: building coalitional power for labour rights between unions and (consumption-based) social movement organisations. Organization 22:720–740
Russell M (2016) Cost is biggest barrier to Bangladesh RMG remediation. Just-style. http://www.just-style.com/analysis/cost-is-biggest-barrier-to-bangladesh-rmg-remediation_id128181.aspx. Accessed 21 May 2017
Rutledge B (2016) Cambodia, Myanmar and the EU's 'Everything but Arms' trade deal. http://www.ethicaltrade.org/blog/cambodia-myanmar-and-eus-everything-arms-trade-deal. Accessed 24 Apr 2017
Steering Committee Members (2017) http://bangladeshaccord.org/governance/. Accessed 24 Apr 2017
The Daily Star (2016) No accord, alliance after 2018: commerce boss. http://www.thedailystar.net/business/no-accord-alliance-after-2018-commerce-boss-1210123. Accessed 21 May 2017
Trade Information (2011) Bangladesh garment manufacturers and exporters association. http://www.bgmea.com.bd/home/pages/TradeInformation. Accessed 8 Jul 2016
Van Heerden A (2010) Making global labor fair. TED Talk. https://www.ted.com/talks/auret_van_heerden_making_global_labor_fair/transcript?language=en. Accessed 21 May 2017
Van Heerden A, Baumann D (2005) Can corporations protect workers' rights? Int Cent Trade Union Rights 12(1):3–4. http://www.jstor.org/stable/41936260
Von Drehle D (2003) Triangle: the fire that changed America. Atlantic Monthly Press, p 40
Zadek S (2004) The path to corporate responsibility. Available via Harvard Business Review. https://hbr.org/2004/12/the-path-to-corporate-responsibility/. Accessed 24 Apr 2017

Part III
Management Approaches and Business Cases

Human Rights in Global Supply Chains

Reflections of Tchibo: A Learning Journey on What Drives Change Towards Ensuring Human Rights of Workers in Global Supply Chains

Nanda Bergstein

1 Introduction

These days, browsing through a daily newspaper seems equivalent to reading excerpts from a horror novel. The news confronting one: terrorism, climate change, exploitation of people in production, violence, fear, and poverty seems to be the order of the day. The world seems to be spinning in the completely wrong direction. A significant part of this scene is related to the effects of globalization and global consumption, areas in which I am strongly involved through the nature of my work. The negative impacts of global consumption and manufacturing, particularly in developing countries, are blatant and there is an imperative need for change, to not jeopardize the future of present and coming generations. Businesses are key to this process.

This paper shares strategies for change in relation to human rights in global supply chains. I am writing from the standpoint of a Corporate Responsibility manager with 10 years of experience in working, in a variety of ways, on human rights and sustainability in global supply chains. My career started out as project manager for a grassroots development NGO in South India related to young girls on reproductive rights and life skills in a financially deprived rural area. After that I joined a consultancy and supported companies and suppliers to install social management systems for better working conditions in factories. Since 2007 I have been an employee of the German coffee roaster and non-food retailer Tchibo where I am responsible for the sustainability and corporate responsibility activities of the company together with my team.

Tchibo is pursuing to become a 100% sustainable business as part of its corporate strategy (Tchibo 2017). The belief that good business is responsible business is

N. Bergstein (✉)
Tchibo GmbH, Hamburg, Germany
e-mail: nanda.bergstein@tchibo.de

S. M. Jastram, A.-M. Schneider (eds.), *Sustainable Fashion*, Management for Professionals, https://doi.org/10.1007/978-3-319-74367-7_5

deeply rooted in the mind-set of the company, which is family-owned and has a strong set of values. We understand that our future business relies on sustaining the raw materials we use in our products—coffee, cotton and wood, amongst others—and establishing fair conditions in manufacturing for the viability of a healthy and motivated workforce.

"100% sustainability" may sound unreachable. However, for us, 100% sustainability means that full social and environmental responsibility pertaining to our business activities is non-negotiable. Questions around feasibility and ambition are not reasons to water down our expectations. On the contrary, this ambitious goal serves as the fuel to propel us into innovation and to strive for new approaches and solutions. With an eye on the present state of the world—there is no alternative to sustainability.

2 Designing Our Human Rights Programme

The Tchibo Human Rights programme is designed around the core conventions of the ILO, the International Human Rights Conventions along with the UN Guiding Principles on Business and Human Rights (Tchibo Kodex 2017).

When we started to build our implementation programme for the non-food business in 2006, we were floundering in unchartered waters. Despite 15 years of CSR industry to that date, there was a scarcity of effective solutions to build upon. Already then it had become clear to us that social auditing was not the answer to drive change at a factory level. The concept of social auditing had been derived from quality management and had adopted the idea that if there was a clear standard with clear criteria of what compliance meant, its implementation could be assured through external monitoring. That logic, however, had a serious flaw. It was based on the assumption that human behaviour could be audited and thereby controlled within the short time span of the 1 or 2 days which an auditor visited a factory. Of course, non-compliances like missing fire exits could be detected, but what about discrimination or sexual harassment? And after concerns were identified, was it sufficient impetus to actual change? Behavioural changes do not occur simply because of a tick in the box on the auditor's sheet (Clean Clothes Campaign 2005). Auditing clearly was not the solution.

Our strategy design was based on studying the root causes for the lack of progress. In our view one important obstacle was the strongly conflicted relation between the workforce and management leading to severe gaps in communication and mutual understanding. Differences in terms of gender, ethnicity, and class, amongst others, made workers inferior in the perception of management. Workers were viewed as a cost for production (and reduced margins), not as creators of value, and certainly not as human beings. In a context of the strong competitive pressures factories felt, one could argue that this perception unconsciously justified putting more and more pressure on the workers. This form of marginalization was exacerbated through a lack of enforcement of national labour laws in many production countries, so that workers hardly receive any protection.

Our human rights programme intended to go to the core of the tension, namely the social relation between managers and workers. Our idea was to deconstruct the construed gaps between workers and managers and to facilitate a further step with the goal of bringing ambiguities together: what if it was possible to protect human rights and at the same time contribute to a sustainable business development of the factories? What if there were win-win situations for both sides, enabling the implementation of human rights? With these questions in mind we created the WE Programme.

3 The WE Programme

WE stands for Worldwide Enhancement of Social Quality. It is a comprehensive supplier training programme which aims at improving working conditions and compliance with the Tchibo Social and Environmental Code of Conduct. Business aspects such as quality, productivity, and efficiency are addressed to the extent that they correlate with human rights, such as working hours or pay.

At the heart of the methodology is dialogue. It is participatory, involving both workers and managers and, where applicable, buyers, to create solutions for human rights challenges. An important part of the process is conflict mediation, which is facilitated by local facilitators specifically trained for this task.

WE was developed in a pilot project between Tchibo GmbH and the Deutsche Gesellschaft für Internationale Zusammenarbeit (GIZ) GmbH in form of a public private partnership. The pilot was co-financed by the German Federal Ministry for Economic Cooperation and Development (BMZ) between 2007 and 2011 (WE Social Quality 2017). From 2007 to 2010 the pilot ran in 40 Tchibo factories located in Bangladesh, China, and Thailand.

The results of the pilot were compelling and showed impact far beyond what was possible through classic auditing and management training (WE Impact Assessment). Convinced, the Board decided to scale the WE Programme to all strategic factories in the producer portfolio in 2011. By the end of 2017, 360 factories had been integrated in total. Approximately 75% of products in our stores are produced in WE factories or in low risk countries (mainly EU). Today, the programme is running in eleven countries: Bangladesh, Cambodia, China, Ethiopia, India, Laos, Myanmar, Pakistan, Thailand, Turkey and Vietnam.

4 What Makes WE Special?

The core of WE is to create solutions for human rights in factories through constructive dialogue between the workforce and the management, as well as between suppliers and buyers. For long-lasting change we envision dialogue to be deeply anchored in businesses and applied as business process, regardless of which retailer sources at a factory. It has to be lived from the inside out.

To enable the parties to engage in dialogue they receive training in communication skills and solution-oriented action planning techniques. Over the course of 2 years per factory we organize space for dialogue in workshops and factory visits and provide the workers and managers with specialist facilitators to help them engage with each other. Whether or not each training participant makes use of this space can only be his/her individual decision. But when they seize the opportunity, they step into their individual power and become part of a transformational process in the factories. In this connection, I have experienced many conversations and inspiring moments in which worker representatives shared with me how proud they were to be able to achieve better working conditions for themselves and their co-workers.

5 The Dialogue Process in Practice

It has always been difficult to describe what actually happens in the dialogue process and to put into words its transformational nature. When possible and with the consent of the participants, I have taken stakeholders and buyers with me to the trainings so that they can see and feel for themselves what occurs in this process.

In the following I attempt to draw an image of a dialogue process, which I attended in Bangladesh. The theme was "working hours," a typical issue surfacing often in working life.

So imagine you are sitting in Bangladesh in a training room with 30 managers and 30 worker representatives from ten factories. At the beginning of the workshop a training on legal requirements around working hours has taken place and afterwards the two parties have been separately working on the topic for 60 min. They have brainstormed the challenges they are experiencing concerning excessive overtime. As the groups join together, there is a buzz of excitement in the room. Which group will have come up with more issues? Who will present better? But there is also tension. Workers are asking themselves whether they will truly be heard. Managers are wondering whether they will face pushback if they share their perspectives. Even though they have the hierarchical "power" the sheer volume of workers in a factory can intimidate them.

In this dialogue round, the workers present first. A young woman stands up. Her name is Asma. She is wearing a beautiful red dress, handpicked by her family because of the special honor to be participating in this forum. This is the dress she wore at her sister's wedding.

Her voice is soft at first, a bit shaky. She looks at me, representative of Tchibo—the client. My reactions as client are relevant to the group and I am aware of the responsibility I carry. I smile at her in encouragement, and she straightens her back and begins:

> This is what we workers have to say about excessive overtime: When we have to work too long, we become thoroughly exhausted. Especially as women, many of us have three jobs. We have to work in the factory, then we go home and have to take care of our husbands and our children. Before we travel to the factory, we wake up very early to cook them meals and clean the house. When we work late, we come home very late, and then we still have to clean the dishes and prepare everything for the next day. We are afraid when we have to travel with the bus at late hours. It's not safe for women. The men edge too close to us and we feel uncomfortable.
>
> Another problem is that we get tired. We become slower at work and we make mistakes. Then we get pressure from our supervisors and that makes us feel bad. We worry about keeping our jobs. But we need them. We also need to work overtime. We need the money to be able to support ourselves and our families

I remember being very impressed by the clear and strong way Asma represented her peers' opinions. Just to set the record straight. How many people in Europe or the US would have the courage to express this or other concerns to their managers right in front of a large group? Of course, the process is designed to ensure that those who voice their opinions feel safe and are well protected. Nevertheless, allowing for such vulnerability and openness is courageous in any context.

This holds true for the managers as well, who, for the first time have to confront the workers. They feel equally vulnerable despite their "superior" role. Owning up to non-compliances towards the client can feel risky for them. For decades they have been incentivized to demonstrate compliance so as not to lose any orders. Double bookkeeping and bribing of auditors has become a common practice because of this, and now they are facing a client who wants to see the truth and the actual problems, even if this means owning up to severe non-compliances (Clean Clothes Campaign 2005).

Our policy differs in that we have committed ourselves to work with our WE factories towards compliance, on the condition that there is transparency regarding the challenges and under the condition that we can see real progress in a reasonable period of time. We believe it is better to address and tackle a problem rather than pretending that it does not exist, because that is an impediment to change. Though gaining the trust of the management to be truly open takes some time and requires patient, respectful communication.

In the managers' group, Shekhar represents his colleagues:

> Excessive overtime hours are also a problem for us. We don't get enough time for our families, especially to see our kids. And overtime doesn't really help because we can see that the quality of the products goes down. Our rejection rates are high. Also, we have to pay the overtime premium, so this is much more expensive. But we don't have a choice, because we have to ship our goods on time; often there are delays by the customers or the raw material is late or we overbook production. Then overtime has to happen. We cannot risk disappointing

our customers and losing orders or pay penalties. But if we had a choice, we wouldn't want it.

By the time the managers have finished, the energy has changed in the room. It has become clearer and brighter. It is as if both parties are really looking at each other face to face for the first time. The anonymity has lifted.

What has happened here? Both parties have opened up and shared their concerns and worries. Both parties have also shown that they equally have a problem with the issue. This creates connection between them, more humaneness and the common goal of finding ways to reduce excessive overtime by considering the needs of both parties.

As a next step in the dialogue process, the workers and managers jointly begin to work on understanding the root causes of the problems they have identified. What is important here: they look at the real issues in the factories so as to develop meaningful solutions.

In the case of overtime, there are usually three big root causes: low wages creating the need for workers to work overtime, unprofessional production and raw material planning, hampering the production flow, as well as last minute changes through customers or delays in order approvals, disturbing the production planning.

The next step is deciding which root causes to work on and how. Here the facilitators play an important role. Their job is to help the factories identify the appropriate aspects, i.e. those they can really tackle and influence.

There is a human tendency to blame difficulties on the one who is not in the room. Often customer behaviour is lengthily discussed, but a factory cannot make an action plan for its customers. This is not to say that clients don't have to change their purchasing behaviour. At Tchibo, for example, as a response to the feedback from our factories we have set up a separate project on fair purchasing practices in our business. However, we also want to support the factories to move out of a victim position into one of empowerment: away from feeling that there are "no options" and "nothing will ever change" towards focusing on those areas with influence. Because this model of thinking will create the greatest independence and impact for the factories.

In this workshop the group decides to brainstorm solutions for the following question: How can we reduce working hours, while maintaining production and increasing wages? As potential solutions they look at creating efficiencies, reducing rejection rates, planning raw material better, optimizing quality control, improving production flow, and skills trainings for workers. They also find solutions to protect female workers, in case overtime has to take place. Examples for actions are accompaniment schemes or buying a company bus.

Finally, the workers and managers negotiate an action plan around the designed solutions, which they take back for implementation in their factories. The facilitators visit the factories between the workshops to offer process support and trainings for the larger workforce.

6 Why Workers Are the Key?

At the start of the programme many factories, but also stakeholders, asked why we believed that workers would actually have valuable input to give, considering the often lower literacy and schooling levels. I will not go down the line and question their beliefs that illiteracy implies less able and challenge the biases that lie at the core of these questions. Instead let us look at the reality: workers are the ones who produce the goods, day in and day out. They know when there are blockages in the production flow. They might not be able to use highly sophisticated terms of industrial engineering (who would?); however, they would say: "my eyes hurt because the light is so weak. This is why, I get slower as the day goes by." They might also say: "I always have to wait a long time until I get the next garment to sew. There seems to be a blockage somewhere." Or they might say: "many of the new girls in my line cannot use the machines properly. They are new. This is slowing us down. Or: my sewing machine is dripping of oil. This is staining the product." But they are rarely asked about their thoughts. They are rarely heard. They rarely feel respected. Why should they share their knowledge? Why would they put themselves into a situation in which they might be criticized or harassed for speaking out, even if it were for the benefit of the company?

Workers do have something to say, to share, and to contribute. And they will do so, if they are respected, listened to and their efforts appreciated through tangible improvements of their working conditions.

What is striking is that sometimes the solutions are utterly simple but happily effective. For example, the suggestion of a worker to improve the lighting in a sewing line increased overall efficiency by 30%. And this is where the magic happens, because suddenly managers see for themselves that integrating and respecting workers makes total business sense.

7 Impact

WE has generated significant change in many areas which are relevant to workers and managers, as evaluated after the pilot phase (WE Impact Assessment).

- Worker participation has increased, formal dialogue structures have been built up and worker representatives have been elected.
- On average, wages have increased by 30–50% (but have not yet reached Living Wage standards).
- Overtime hours have been reduced.
- Welfare programmes have been set up at factories.
- Health and Safety management systems have been introduced
- Efficiency and quality has improved.

As there is a clear methodology and approach behind generating these results, the approach can be replicated into other supply chains and business models.

Topics with more systemic root causes, which go beyond the factory boundaries, are more difficult to tackle, e.g. full living wage. For this we are working on complementary solutions that I will describe in a later section.

8 Learnings on Success Factors

A number of factors enhance the success of the dialogue-oriented supplier training.

- *Securing management commitment*: From the start of the training, we involve factory owners and top management to ensure their support. We invite them to the training activities and ask them to be present at WE workshops for timely decision-making on action plans. Where necessary, the Tchibo buying department gets involved so as to secure their commitment. This sometimes requires detailed conversations and they are not always successful. When top management consistently fails to show commitment, we define this as a difference in values which can even lead to ending the business partnership.
- *Locally adapted and culturally sensitive training while maintaining international labour norms and human rights*: The WE trainings are adapted to the individual conditions of the programme countries. This does not mean, however, that international social standards are watered down. The stricter norm applies according to our buying contracts. But we try to bring across the message in a way which can be embraced by the participants. For this we strongly build on our local facilitators. We acknowledge and appreciate that they come from a mix of experience. We make teams of people with NGO grassroots, union, and consultancy backgrounds, so that they are able to address both the workers as well as the top management and produce optimal results. The challenge which we have encountered here is to identify the right facilitators for the job. It requires a lot of skill and experience to identify, train, and become a WE facilitator. We continuously invest to upgrade our standards and innovate our training methodology.
- *Facilitating participation and legitimate worker representation*: In WE we work intensively on creating permanent worker representation structures in the factories, optimally involving unions when existent. Democratically elected worker representatives can best reflect the opinion of the entire workforce. Developed solutions will represent the interests of the workers more holistically and the workforce will be more satisfied with the negotiated outcomes. Worker representatives with a union background tend to be more effective because they have a support system outside of the factory.

  In many factories, worker representation is not functional. This is why we design a motivational process into the WE training, in order to demonstrate to managers the positive implications of involving workers and building representation structures. Here we make sure to clearly abide by Art. 2 of the ILO

Convention 98 on the Right to Organise and Collective Bargaining (1949)[1]. It is our responsibility to open a space for worker organization and ensure that workers know their right to freedom of association, but it is the workers' responsibility and decision whether or not to exercise their freedom of choice. Influence in any direction can weaken the position of the worker representatives, as they become management influenced, undermining their credibility vis-à-vis the workers. This is why we also signed a Global Framework Agreement with the Global Union INDUSTRIALL in September 2016 in order to strengthen the positive effects of freedom of association and collective bargaining on workers and business through a better integration of unions.

– *Ownership for the beneficiaries—putting them centre stage*: WE empowers workers and managers to shape change according to their needs and wishes. The goal is for them to have full ownership of the process. "Ownership" means having the freedom of decision: whether or not to participate in the programme, which topics to tackle and when. The only frames, which we set, are the labour standards and human rights within our Social and Environmental Code of Conduct which refer to the UN Guiding Principles on Business and Human Rights (UNGP) as well as the International Human Rights Covenants. All of them have to be addressed, but the concrete focus is dependent on the joint decision of the workers and managers. The facilitators guide the workshop participants towards creating their own personal breakthroughs, augmenting ambition and deeper implementation, with great sensitivity, credibly ensuring that workers' needs are considered.

  Ownership for the beneficiaries means letting go of the outcome for Tchibo. This felt unsettling in the beginning, as results are imperative for our own values and for our stakeholders. It seemed counter-intuitive to consciously let go of our expectations in the discourse with the training participants. But here we wanted to learn from the past: imposing our Code of Conduct in a top-down manner had not gotten us very far, as previously discussed in the section on audits. Opening space for the participants and keeping the decision making power with them, however, generated more change than we would have imagined possible, so that we have kept this practice until today.

  The key lies in allowing the participants to experience the positive feeling of success. Showing them that they are capable of developing their own solutions and implementing them boosts their confidence and makes them proud. In the beginning, the implementation decisions tend to be small steps, and not at a level of ambition we would expect. But it is understandable: the workers and managers

[1](a) Workers' and employers' organisations shall enjoy adequate protection against any acts of interference by each other or each other's agents or members in their establishment, functioning or administration.

(b) In particular, acts which are designed to promote the establishment of workers' organisations under the domination of employers or employers' organisations, or to support workers' organisations by financial or other means, with the object of placing such organisations under the control of employers or employers' organisations, shall be deemed to constitute acts of interference within the meaning of this Article. ROCBG (1949).

are testing the waters and trying to understand whether they can trust us and trust each other. With first implementation results they realize that they can do it, and from action plan to action plan the decisions become more courageous. This triggers change from the inside out.

- *Peer learning*: Exchange between the suppliers during the workshops provides additional stimulus. They present the progress they have made to each other and in doing so create a constructive atmosphere of competition and discuss challenges and offer solutions to each other.
- *Setting boundaries if there is no progress:* In WE, suppliers receive ample freedom to work on their issues. We do not set content goals, but we communicate the expectation that there has to be clear and tangible progress within the course of the programme, and also on the topics which matter most to us, e.g. wages, freedom of association or discrimination (duration of 2 years). Suppliers receive time to get on track. If they show massive resistance to change over the course of the programme, however, the sourcing department will review whether such an organisation fits the principles and values of Tchibo and in severe cases will cut business ties.

Overall, we have started to apply these principles to other sustainability programmes which we run as well and can see that they enhance the effectiveness of our work. In a nutshell, the parameters described above create a space which is designed in such a way that the workers and managers can have a voice and develop solutions with self-determination. The process to get to that space, in which power, gender, and cultural hierarchies are balanced out, takes time however, sometimes up to a year. One could say that the downside of the WE approach is the need for patience and continuous dialogue with the factories. In essence, we as sustainability community have to realize that there is no quick fix. But the investment is definitely worth the results.

9 Moving Beyond Single Supply Chain Approaches to Address More Sectoral and Systemic Issues

WE is greatly successful with all issues which are within the realm of the individual influence of factories and Tchibo as a brand. But we have had to acknowledge the boundaries of this approach. There are topics which we have tried to solve unilaterally to no avail, simply because there were too many external factors of influence related to the dynamics of globalization and political frameworks in the sourcing countries.

We had to face up to the realization that real change in our individual supply chain requires looking at and influencing these external factors. Therefore, after 5 years into the WE programme, we started promoting and engaging in new formats of collaboration to create a shift for the entire industry, thereby positively affecting our own supply chains.

10 Sectoral Engagement: The Bangladesh Accord on Fire and Building Safety

The first opportunity arose in 2012 when the ITGLWF (now a part of INDUSTRIALL) and the Clean Clothes Campaign invited Tchibo to join in negotiating the Memorandum of Understanding on Fire Safety (MoU), which they had begun developing with Philip van Heusen (PvH), (Clean Clothes Campaign 2013: 2). The MoU's goal was to promote fire safety in Bangladesh at a systemic level including the following elements: bringing in the missing expert knowledge on fire and building safety into the country, setting up fully independent factory inspections, creating transparency for the public over inspection findings, and strengthening the involvement of workers, especially unions. Especially the latter was extremely important to us and decisive for our engagement. Previous fires and tragedies (e.g. the Spectrum collapse in 2005) had shown that workers had known that something was wrong in the factory, but that they were driven back to their workplaces when they spoke up. Deaths could have been prevented, had they had the power to refuse unsafe work. Next to the involvement of unions we also joined the MoU because it was legally binding. Our intention was to create a framework which would hold us and others accountable, especially in situations where safety and economic interests could be in conflict, e.g. potential cases of factory closure or ensuring the financial underpinning of remediation. These would be the cases that would test the signatories' commitment. A legally binding framework would provide the backbone to implement necessary steps.

Tchibo joined the MoU as part of our due diligence strategy. We knew that despite our investments (e.g. mandating German engineers to inspect our factories) we would never be fully able to guarantee fire and building safety, unless there was a sector solution.

Originally, the MoU was set up to involve four to five large brands jointly with INDUSTRIALL and the Clean Clothes Campaign. For several months the signatories (Tchibo & PvH) jointly tried to find further partners, but progress stalled. At that time legal collaboration with the labor side was met with hesitation due to a lack of trust that a true partnership could be formed with the "other side".

When Rana Plaza, happened, however, INDUSTRIALL used the opportunity to align all big brands and retailers, and the MoU transitioned into the Bangladesh Fire and Building Safety Accord.

Today more than 200 companies are members of the Accord. The initiative has contributed greatly to a safer factory landscape in Bangladesh. Needless to say there is still a lot of work to be done which is why we have signed the extension of the Accord. I would argue that next to the impact on the ground, the Bangladesh Accord has demonstrated that an industry-wide approach can drive change in a country on a specific topic.

11 Tackling Global Issues: Act on Living Wages

With the positive experience of the Bangladesh Accord to back us, a group of brands came together from 2014 onwards to develop effective solutions for living wage. The Accord had shown that sector collaboration on implementation was possible, and within the Ethical Trading Initiative (ETI) and the Accord a group of companies, including Tchibo, H&M, C&A, Inditex, Next, and Primark, amongst others, had evolved which was interested in driving into even deeper and more ambitious change.

Tchibo decided to engage in the group since single supply chain approaches like WE could achieve tangible wage increases, but apart from China, we were not meeting a living wage level, e.g. benchmarked against the Asia Floorwage or the SA8000 standard. Tchibo had already tested various models with key supplier factories in different countries, working on living wage in combination with worker participation, industrial engineering, and higher buying prices. We saw progress under very restricted conditions, i.e. when we were almost the only buyer in a factory and when a strong union was involved. This, however, is not the norm in global supply chains: Tchibo offers goods from 75% of Non Food categories. Because of that, we source our products from around 700 factories. We very rarely have over 50% buying share, which—by the way—also creates too much dependency for factories and is not wise for a healthy business. Against that background, the models we had tested were not scalable for us and presumably also for most companies in the industry.

Additional challenges added to the restrictions of single supply chain approaches:

- Peer pressure on suppliers not to raise wages to an extent that might cause desirability leading into strikes in other factories,
- A lack of unionization of factories, which meant that there was no guarantee that buying price increases would be passed on to the workers,
- Competition that was driving the entire industry to be as cost competitive as possible,
- Consumers who had been educated to always expect a "cheaper deal" and who have not been willing to share the added costs of wages and sustainability until now. Without an equal level playing field requiring all companies to incorporate higher wages, more progressive companies could be challenged in their competitiveness when reflecting living wages in sales prices.

Our discussions as a group led us to the conclusion that we had to look at the system that was driving down wages. We needed to find a mechanism that defined wage levels as a non-negotiable, a mechanism that enabled progress towards a living wage, and a mechanism that ensured that higher wages would actually reach the workers. Also we were clear that brands' and retailers' purchasing practices had to be part of the equation. The way we buy has to incentivise the payment of a living wage.

We also requested a partnership with the global union INDUSTRIALL to work on the situation in a joint endeavour. The ACT Initiative was set up and a Memorandum of Understanding was negotiated to formalize our relationship. Further, as companies do not have a formal representation structure we sought out facilitation to help us achieve joint positions and be able to work with INDUSTRIALL in a coordinated manner.

ACT aims to improve wages in the industry by establishing industry collective bargaining in key garment and textile sourcing countries, supported by world class manufacturing standards and responsible purchasing practices of suppliers, brands, and retailers.

The reason why we believe that industry collective bargaining is central, is that it brings together national representatives of manufacturers and workers—namely employer associations and trade unions—to negotiate and agree on wages and conditions applicable to an entire industry sector within that country.

The collective agreements that result are legally binding and enforceable. Industry-wide agreements set a benchmark that applies to all manufacturers, while still allowing for individual manufacturers to offer higher pay and better conditions (ACT 2015).

As I write we are in the phase of formalizing the ACT initiative and planning implementation. We believe that this is the right way forward, but only history will tell whether or not we were successful. We are aware of the responsibility that we have as a community to make this work. It is time for the next level of change, but as in the past 10 years of our individual journey as a company, we know we individually and we as a group will learn a lot along the way and will most make many mistakes to then adapt and redesign our approaches.

12 Conclusion

In this paper, I have described strategies on how to implement human rights in global supply chains, working in individual supply chains, as well as on a country and global level. Crucial to success is creating partnerships with the labor side, strengthening worker participation and the voice of unions in the factories, as well as redesigning the purchasing practices of brands, retailers and suppliers. Achieving this will take time and further innovation. There is no quick fix and it is best to acknowledge that, without making this an excuse for non-action or non-participation. On the contrary, I believe this should be fuel to step up our game.

As a CSR community, I hope that we will constantly have the courage to question our strategies and approaches, to understand what works and what does not work and also why. I hope we will be responsible enough to acknowledge that this is a learning journey that entails the sharing of lessons learned, preserving what is effective, but also eschewing models or approaches which have fallen short of impact, in order to create space for innovation, as difficult as this might be given the personal investment of energy and time. As an industry we owe this to thc people producing our goods, to our consumers, future generations, and to ourselves.

I have learned to anchor myself deeply into the “WHY” of my work. Whenever I think that the journey is getting too hard and I feel like giving up, I remember Asma from the WE workshop and her courage to speak up. She probably does not know it, but her speech has been an impetus to the work of my colleagues and myself. Her picture hangs in our department as a symbol of the effect that one individual can have on the entire supply chain.

Asma inspires us to use our own voices and to continue to find and develop solutions for global change.

References

ACT (2015) ACT factsheet. http://www.industriall-union.org/sites/default/files/uploads/documents/2015/Cambodia/act_factsheet2.pdf. Accessed 15 June 2017

Clean Clothes Campaign (2005) Looking for a quick fix. https://archive.cleanclothes.org/documents/05-quick_fix.pdf. Accessed 15 June 2017

Clean Clothes Campaign (2013) The history behind the Bangladesh fire and safety accord. http://www.cleanclothes.org/resources/background/history-bangladesh-safety-accord. Accessed 15 June 2017

ROCBG (1949) Right to organise and collective bargaining convention 1949. http://www.ilo.org/dyn/normlex/en/f?p=NORMLEXPUB:12100:0::NO:12100:P12100_INSTRUMENT_ID:312243:NO. Accessed 15 June 2017

Tchibo (2017) http://www.tchibo.com/nachhaltigkeit. Accessed 15 June 2017

Tchibo Kodex (2017) Verhaltenskodex Zuliefer. http://www.tchibo-nachhaltigkeit.de/servlet/cb/954990/data/-/VerhaltenskodexZuliefererEnglisch.pdf. Acessed 15 June 2017

WE Impact Assessment. https://www.we-socialquality.com/WE-Specificity/First-results.aspx?l=1. Accessed 15 June 2017

WE Social Quality (2017) Worldwide enhancement of social quality. www.we-socialquality.com. Accessed 19 June 2015.

Creating Sustainable Business Partnerships at H&M Through Collaboration

Hendrik Heuermann

1 Introduction: The H&M Way of Production

At H&M we want to make fashion more sustainable, affordable, and attractive for as many people as possible. We have known the pros and cons of worldwide garment production since we were founded in 1947. It is a labor-intensive industry that creates millions of jobs globally. The textile industry has proven to be an accelerator of industrial development and economic growth in manufacturing countries in several cases (ILO 2014a). At the same time, human rights standards and environmental awareness are less developed in many of these countries. Within the industry, it is important to ensure that growth and newly created jobs actually affect people and communities in a way that brings about positive change.

As much as H&M is part of a global industry with a large number of direct competitors, our way of working in the production phase is different than that of many others. H&M currently has 21 production offices worldwide to locally organize production. Many other brands use textile agencies, instead, to handle the largest part of their value chain. In this case, the brand's core design team hands over a sample of a product to a textile agency which then arranges the production, chooses suppliers, checks the quality, arranges transport, etc. H&M instead invested in a local presence to organize production on site in order to maintain close relationships with our business partners. Approximately 3000 people worldwide handle H&M's global production.

The focus is on building strong, long term business partner relationships that are characterized by trust and transparency.

H. Heuermann (✉)
Sustainability Manager Germany, H&M, Zürich, Switzerland
e-mail: Hendrik.Heuermann@hm.com

S. M. Jastram, A.-M. Schneider (eds.), *Sustainable Fashion*, Management for Professionals, https://doi.org/10.1007/978-3-319-74367-7_6

With 220 full-time employees working exclusively on sustainability in production offices, the Swedish head office and sales markets, we can have the most direct influence on our partners and work with them towards constant improvement.

2 Improving Strategic Partnerships

We believe in the financial payoff of working directly with our suppliers. Not only is it more responsible to have a good overview of your direct business partners, it's also a wise business decision.

As we grow together with our partners, we benefit from more stable and reliable business partners with an increased capacity.

The H&M Group works with 1826 first tier factories. Before we work with a new factory, we perform a detailed assessment on site to ensure that our minimum requirements are met.

If the facility is approved, the supplier signs our Sustainability Commitment (SC). The SC is shared by all brands within the H&M group, and everyone who has a business relationship with us must sign it if they want to start or continue working with us.

The Sustainability Commitment clearly states that compliance with the law is fundamental (H&M 2016), but we want to reward more aspirational partners with better business in terms of long-term business commitment and growth opportunities. This has proven to be much more effective in terms of sustainability performance.

That is why our SC has two levels of performance: fundamental and aspirational. The fundamental level is the basic compliance level that we expect all business partners to meet. The aspirational level demonstrates where we want to go in the long term through constant improvements and shows that compliance with legal requirements is only the first step in creating a sustainable business.

For example, in addition to setting requirements for wastewater treatment and water quality, the new commitment now takes a stewardship approach and covers water withdrawal and its impact on local communities and the environment. This encourages suppliers to develop their own strategies for tackling water challenges as well as providing incentives to shift impacts from negative to positive (H&M 2016).

In our performance evaluation, we reward business partners that commit to and meet their aspirations. The three major focus areas of our Sustainability Commitment are healthy workplaces, healthy ecosystems and a nimal welfare.

One of the many responsibilities of our local production offices is to monitor the work and the progress of the supplier factories. An audit or inspection of the factory can be used as a tool for this, and for many years, we have used this tool. Despite having a different image in the media, our audits were developed from a compliance based check referring to a code of conduct established in 1997 to a coaching relationship in our Supplier Impact Partnership Programme (SIPP) and the joint Sustainability Commitment (SC) between the supplier and H&M. The SIPP supports suppliers in improving their sustainability performance and helps us monitor and measure suppliers' compliance against the fundamental and aspirational levels and assess the level of implementation of our Sustainability Commitment.

After a facility is approved and the SC is signed, its sustainability performance is then self-assessed annually. Its overall performance and country risk assessments, amongst other indicators, will help us to determine the level of validation that is needed for each factory. For high-risk and new business partners, we perform an onsite check at least once a year. If the business partner is low risk (i.e. they achieved a high score on the audit), a 'desktop validation' is usually sufficient. This involves checking data and paperwork for anomalies and inconsistencies. We assess each business partner's capacity to both manage their sustainability issues and track actual performance using environmental and social sustainability metrics on a three-leveled approach:

- Level 1: Compliance with fundamental requirements (such as legal requirements, the ILO core conventions, and other requirements similar to our previous Code of Conduct).
- Level 2: Existence and quality of management systems and setting of their own sustainability performance goals. Management systems help to sustain compliance and establish a fundamental ground for leading practices and reaching aspirational goals.
- Level 3: Setting of and performance towards aspirational targets based on defined Key Performance Indicators (KPIs) such as energy reduction or community impacts.

We combine our suppliers' performance across these three levels in our new Sustainability Index, which fully replaced our Code of Conduct Index (IcoC) during 2016.

This new assessment method is based on the Higg Index and some complementary questions and KPIs developed by H&M. The Higg Index is a holistic measurement standard. It was developed by a broad range of brands (including us), suppliers, NGOs, academia, and many more through the Sustainable Apparel Coalition (SAC). The aim of the Higg Index is to create alignment, greater transparency, and better sustainability performance across the apparel and footwear industry (The Higg Index 2017).

The ambition with SIPP is the desire to change from a one-way compliance communication (audits mostly) to a two-way dialogue about business improvement.

We conducted our first SIPP assessments during the second half of 2015, which covered all of our first-tier supplier factories. In the future, we will also include processing factories and nominated second-tier factories, such as fabric and yarn mills, and suppliers of non-commercial goods, such as advertising materials. Our partners have the responsibility to constantly monitor their sustainability performance themselves and track their progress in an online tool.

With greater insight into the data, we can focus our efforts on our capacity-building programmes. These programmes represent all the supplier improvement activities within our road maps, such as water, climate, energy, chemical management, industrial relations (i. e. the relationship between employers' and workers' representatives), and fair living wages.

One of the outputs of the assessment through SIPP is our Sustainability Index (SI). Each factory is given an SI score between 0 and 100 based on the Higg facility module and our own KPIs and is then ranked according to its environmental and social sustainability performance.[1] Ninety-six percent of the factories were assessed before the first order placement; we performed 1.194 minimum requirement assessments, 459 desktop validations, and 1.247 on-site validations (H&M 2017: 100).

Only those of our direct or "first tier" suppliers who demonstrate the best sustainability performance can become strategic partners.

This type of partnership represents a win-win situation, for instance, by securing stable business through long-term capacity planning up to 3 years in advance for "Gold suppliers" and 5 years for "Platinum suppliers".

In turn, the close integration of sustainability performance into our purchasing routines also allows us to reduce orders in the case where a supplier shows unwillingness to work with us on improvements. We are clearly committed to working together on improvements rather than terminating relationships. However, if a supplier repeatedly does not comply with our minimum requirements, we will, for example, reduce order volumes as a warning. As a last resort, we terminate our business relationship.

With a yearly global supplier survey and meetings with all of our vendors, we can learn what we need to improve. One important aspect is their perception of us as a fair business partner. Since 2014, suppliers' perception of H&M as a fair business partner has increased from 76.2% to 83% (H&M 2017: 123) (Fig. 1).

We clearly benefit from good and close relationships with our suppliers—both from a business perspective as well as a sustainability perspective. Further, we are confident that collaboration should not end at the factory doors.

Complex wage structures in production countries are a typical challenge of the industry where working with other brands, unions, and local governments is part of the solution.

3 Complex Wage Structures as a Core Challenge

Workers in a textile factory usually earn the same money regardless of for which brand they make garments. Sustainable success would mean that the wages do not depend on which brands the factories produce (Holdcroft 2015).

There is absolutely no question that every garment worker should earn enough to have a decent living. In order to ensure this throughout our industry and in all of our sourcing markets, many parties must still take a big step. Our vision is to have all

[1] At the time of the publication of this book, we were currently updating this strategy and therefore are not able to share aggregated scores yet.

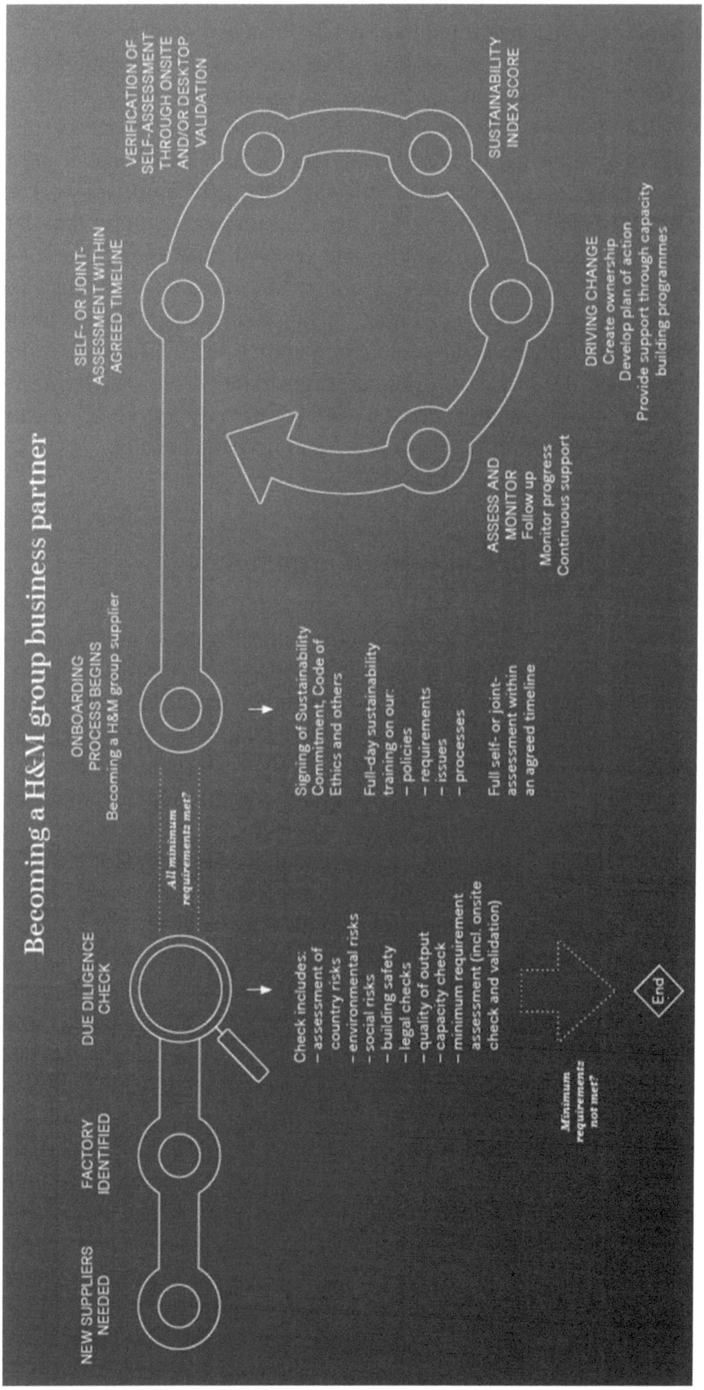

Fig. 1 Process of becoming a H&M group business partner (H&M 2017)

suppliers making our products pay their workers a fair living wage so that they can cover their families' basic needs and have discretionary income.

A minimum wage is the lowest possible wage that workers are entitled to receive. It is defined by local law and can be effective for certain sectors or a whole country regardless of the industry. Paying the legal minimum wage is one of the core, non-negotiable requirements for H&M suppliers. In fact, only a small portion of our suppliers' workers are paid the minimum wage; the vast majority of workers earn more (Fig. 2). However, a minimum wage does not necessarily mean that the basic needs of the employees and their families are covered. That is why we cooperate with many others to strive to create a fair living wage. In our Sustainability Commitment we have clearly stated that "a fair living wage should always be enough to meet the basic needs of employees and their families, and provide some discretionary income" (H&M 2016). Further it should reflect the local living costs of workers and their families, and we are convinced that it needs to be negotiated with democratically elected worker representatives and revised annually.

4 Why We Believe in Collective Bargaining

While many different actors want to contribute to higher wages in textile production, the path to that goal can be quite different as seen in discussions with non-governmental organizations (NGOs), for example, the Clean Clothes Campaign (Clean Clothes Campaign 2013). Among others, these organizations have called for companies such as H&M to adopt a concrete number as a definition for fair living wages. Although we value exchange with NGOs on this topic as well as other subjects, we believe in a different approach. According to the International Labour Organization (ILO) and global unions, there is no universal methodology on how to calculate a living wage. Instead, they stress the importance of promoting freedom of association and collective bargaining as necessary for workers and employers to negotiate wages and conditions of work (Anker 2011; Holdcroft 2015; ILO 2014b).

This is underlined by Jenny Holdcroft, Assistant General Secretary IndustriALL, stating that "Even if buyers increase the prices they pay, without collective bargaining in place there is no guarantee that the increases will be passed on to workers. Furthermore, most suppliers have multiple buyers, all of whom negotiate prices with them individually" (Holdcroft 2015, 101).

We share this view and that is why we focus on improved industrial relations, meaning a good dialogue between employers' and workers' representations. We are convinced that collective bargaining is the best way to define wages. In many markets, however, workers have limited possibilities to negotiate wages collectively. In cases where legal minimum wages or the systems to define them are not yet sufficient, we need to find interim solutions that can ensure that individual factories can pay their workers fair living wages without putting their business and jobs at risk. Income equality also needs to be seen in relation to other sectors, and the impact on labor markets and employment must also be considered. It is also important that

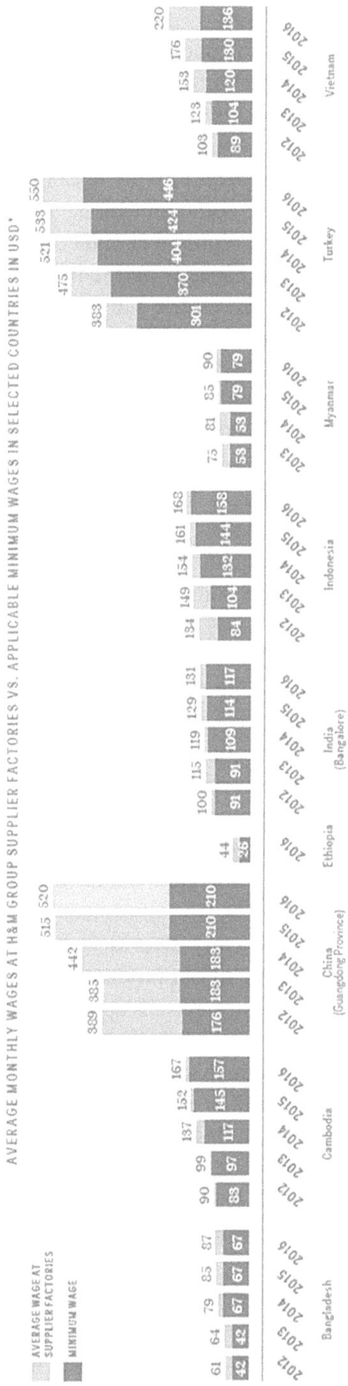

Fig. 2 Monthly average wages at H&M supplier factories (H&M 2017)

governments feel confident that buyers would not move orders to other countries if minimum wages were to be increased.

There is no doubt that it is our job and obligation as a major brand to constantly monitor and improve our own purchasing practices, in order to ensure that our suppliers can pay their textile workers for the true cost of labor and nobody has to work excessive overtime—one of the industry's most pressing issues in our eyes. The purchasing practices should always provide reasonable lead times, fair pricing, timely payments, and transparent communication. To achieve these objectives, better purchasing planning tools are required to reduce production peaks and thereby help our suppliers better plan their capacity and reduce overtime.

This alone will not be enough to ensure fair living wages, but it will give our suppliers additional confidence in implementing improved pay structures.

Functioning industrial relations and collective bargaining are key to achieve good working conditions. Workers must be free to form or join a trade union and be able to bargain collectively about wages that also reflect their individual skills.

We want to make sure that workers are aware of their rights in their factory. This is why we invest in social dialogue and skills development to ensure that workers are hired and paid according to their skills.

In 2016, we continued to scale up our industrial relations and workplace dialogue programmes, reaching 290 factories compared to 132 in 2015. We also introduced a programme in Ethiopia for six factories, covering 4000 workers. In 2017, we plan to expand this to cover Indonesia, Turkey, and Myanmar.

The H&M group is currently using the Fair Wage method, which aims to create management systems in which workers know what wages they are entitled to (including overtime compensation and benefits), as well as ensuring they are paid according to their skills, experience, and performance.

In 2016, we scaled up the implementation of the Fair Wage method reaching a total of 140 compared to 68 in the previous year. We introduced and implemented the Fair Wage method to three new markets: Indonesia, India, and Turkey. In 2017, we plan to include Myanmar.

It is our goal to have democratically elected worker representatives at supplier factories worldwide producing 50% of our products by 2018. Given widely varying pre-conditions from market to market, we are applying locally-adapted approaches to achieve this. This means that where functional union and collective bargaining systems exist, we are continuing to help strengthen them (for example in Cambodia). Where this is not the case, we will work with our strategic suppliers to develop factory-based worker representation that is democratically-elected.

We are convinced that social dialogues on the factory level and industrial relations on the industry level will lead to better negotiations between workers and management and build one of the foundations of a system where fair living wages are possible.

Our goal is for all workers at H&M supplier factories to have access to relevant skills development programmes—and to have employment and a wage that reflects their skills. As part of this objective, we teamed up with the ILO and the Swedish International Development Cooperation Agency (Sida) in Bangladesh in 2013.

Looking at the scale of our approach, we think we can be proud of what we have achieved—however, we want to change the industry for the better and we can't do it alone.

5 Conclusion: Why We Need to Collaborate

To achieve that systemic change in the global textile industry, we need strong partners. That is why in 2014, we entered into a partnership with the ILO to promote sustainable supply chains in the garment industry on a global, national, and corporate level.

During 2015, we signed a global framework agreement with the global union IndustriALL and the Swedish union IF Metall which were renewed in 2016. We are joining forces to promote sustainable industrial relations and collective bargaining globally and improving the dialogue between employers and employees at H&M supplier factories. Many of the countries where we operate have little experience of positive dialogue with the parties in the labour market. That's why, as part of this agreement, together we are helping suppliers and local trade unions to solve conflicts peacefully, primarily on the factory level where they arise. To support this, we have set up national monitoring committees with two representatives from IndustriALL and H&M respectively in Cambodia, Bangladesh, Myanmar, Turkey, and Indonesia.

To influence the whole industry, we need to work with the brands of which we share factories even though they might be competitors on the shopping streets.

Therefore, we are a founding member and we are actively engaging with other brands through the industry platform ACT (Action, Collaboration, and Transformation) to address the issue of living wages in the textile industry. Fifteen brands and IndustriALL have signed joint principles—a shared approach towards a fair living wage based on responsible purchasing practices, collective bargaining (and freedom of association), and skilled workforce. Some of the participating brands include Tchibo, C&A, Inditex, New Look, Primark, Topshop, Tesco (ACT 2016).

Only in the case where the industry takes action together can the undercutting of standards be prevented and can higher wages and better working conditions become competition neutral—a goal ultimately shared by companies.

References

ACT (2016) ACT initiative factsheet. http://s3-eu-west-1.amazonaws.com/www.ethicaltrade.org.files/shared_resources/act_factsheet.pdf?NOLqDau6XQP.b_YbAuXUs1DAwPS4jPPV. Accessed 27 June 2016

Anker (2011) Estimating a living wage: a methodological review. In: ILO conditions of work and employment series No. 29. http://www.ilo.org/wcmsp5/groups/public/@ed_protect/@protrav/@travail/documents/publication/wcms_162117.pdf. Accessed 27 June 2016

Clean Clothes Campaign (2013) Why an Asia floor wage? http://www.cleanclothes.org/livingwage/what-is-the-asia-floor-wage. Accessed 27 June 2016

H&M (2016) Sustainability commitment. http://sustainability.hm.com/content/dam/hm/about/documents/en/CSR/Sustainability%20Commitment/Sustainability%20Commitment_en.pdf. Accessed 27 June 2016

H&M (2017) The H&M group sustainability report 2016. http://sustainability.hm.com/content/dam/hm/about/documents/en/CSR/2016%20Sustainability%20report/HM_group_SustainabilityReport_2016_FullReport_en.pdf. Accessed 19 June 2017

Holdcroft (2015) Transforming supply chain industrial relations. Intern J Lab Res 7(1–2):95–104

ILO (2014a) Wages and working hours in the textiles, clothing, leather and footwear industries. In: Issues paper for discussion at the global dialogue forum on wages and working hours in the textiles, clothing, leather and footwear industries. http://www.ilo.org/wcmsp5/groups/public/---ed_dialogue/---sector/documents/publication/wcms_300463.pdf. Accessed 27 June 2016

ILO (2014b) Global wage report 2014/15. Wages and income inequality. http://www.ilo.org/wcmsp5/groups/public/---dgreports/---dcomm/---publ/documents/publication/wcms_324678.pdf. Accessed 27 June 2016

The Higg Index (2017) http://apparelcoalition.org/the-higg-index/. Accessed 22 June 2017

Value Chain Impact Measurement at the Otto Group

Andreas Streubig

Sustainability does not just happen by itself. It needs a clear attitude and the necessary structures and processes. The Otto Group has both.

1 Where Have We Come From: And Where Are We Today?

History shows us that the world has always had its visionaries. In the early eighteenth century, Hans Carl von Carlowitz, Chief Mining Official in Saxony and Head of the Mining Authority in Freiberg, was such an inspirational thinker. He was responsible for the entire timber supply for the Electorate of Saxony's mining and metallurgy operations. In the face of shortages in timber supply at the time, he described the fundamentals of future-proof management—with reference to forestry—in his *Sylvicultura oeconomica* (Von Carlowitz 1713). This, in turn, established what we know today as 'sustainability': an approach to managing future outcomes that now extends far beyond the borders of Saxony. Given the knowledge of later generations, some of Carlowitz' decisions and initiatives are not entirely positive. Nonetheless, he has made history. He would, however, never have imagined that the idea he had written down, about working respectfully and carefully with nature, would not be taken seriously over the centuries to come.

The consequences have since become clear. By as early as 2050 our Earth will be home to at least nine billion people and is rapidly approaching the limits of its own potential. If we take a look at the 'planetary boundaries' framework (Steffen et al. 2015) that was first introduced in 2009, we can clearly see that four of the nine indicators have exceeded their safe margins. In this framework, the areas of climate change, global warming, loss of biodiversity, land-use, and the global nitrogen and phosphor cycles are especially critical. This signals the end of a 12,000-year period

A. Streubig (✉)
HUGO BOSS AG, Metzingen, Germany
e-mail: andreas_streubig@hugoboss.com

S. M. Jastram, A.-M. Schneider (eds.), *Sustainable Fashion*, Management for Professionals, https://doi.org/10.1007/978-3-319-74367-7_7

of broad stability. The reasons for this are almost exclusively attributable to human activity. So much so that some scientists are already proclaiming a new geological era, the 'anthropocene', in which the developments and changes on our planet are mainly caused by human beings.

Some things are still in a sorry state in terms of social development. At least some of the UN objectives adopted for the millennium in 2000—such as the intended 50% reduction in world poverty and improved access to clean drinking water for the poorest people in the world—are deemed to have been achieved. But more than three-quarters of the improvements agreed upon at the time could not be realised in the 17 years since the turn of the new century, including a significant reduction in child and maternal mortality and a halt to the spread of HIV.

No-one would dispute that many emerging countries have benefited, and continue to benefit, from the economic development as a result of globalization. If we look at the overall picture, however, an ever-increasing divergence in the distribution of, and participation in, these benefits can be observed. The successes mentioned above in the global fight against poverty and improvement in the living standard of the many go hand-in-hand with even greater 'successes' in increasing the welfare of the few. In the light of a study by Oxfam (Hardoon et al. 2016), the richest 1% of the world population has acquired more property for itself than the remaining 99%. While this development of an increasingly unequal distribution of wealth is not fundamentally new, its scale is quite shocking.

This objective aggravation of global problems brings incalculable risks with it: immediate impacts, ranging from shortages of resources to an increase in catastrophic events, and indirect ones, from a rise in distribution conflicts to turbulence in economic and currency systems etc. From this, a long-term danger develops for business and trade in societies, as most existing value-creation models need stable social systems and infrastructure for them to function, in addition to security of resources and supply.

2 What Role Do Companies Play?

Concerning the movers and shakers in society who can drive forward more sustainable development, there is no getting around companies and the people who manage them. They do not, after all, operate in a vacuum. On the one hand they draw on indispensable environmental and social resources in their value creation and, at the same time, their business activities have various consequences—both positive and negative (the latter perhaps unintentionally but nevertheless consciously). On the other hand, business is a key performer in society: it creates tangible and intangible goods, employs people, and influences political and social development processes. As business plays such a central role, it also has a major influence on its business partners, suppliers and customers, as well as on employees, politicians, and the media—indeed, on all entities that play a key role in the governance and guidance of political and social processes.

As a result, most companies are now fully aware of the demand for a more sustainable and consequently a more responsible business community—and hardly anyone would dispute the necessity for this shift in awareness. But having now considered the 'if', a new question comes into focus in the sustainability discussion. The question of 'what': that is, what is the significance or materiality of individual fields of action. What is to be done? And how urgently? And straight after that we come to the 'how': how should sustainability initiatives be deployed, and where? Because we do not just want to respond effectively (in terms of the challenges addressed) but also efficiently (in terms of a balance between cost and benefit). Intuition and an empirical approach will hardly help us any more. We need to find credible answers if we want a modern sustainability management programme. Getting the necessary transformation process to work in companies and society is, and remains, an act of strength. For this reason, it is good to deploy the limited resources in the right way.

3 How Does Sustainability Management Work Within the Otto Group?

At the Otto Group a management process exists for this, named 'impACT'. It combines three integrated and interacting processes: firstly, the materiality analysis itself (the 'what'), which is based on quantitative and qualitative assessments. In parallel with the dialogue that has long been customary with our company's internal and external stakeholders, for 2 years now we have also been using a tool called 'estell' (Systain 2017), developed by Systain Consulting, a company that is also part of the Otto Group. Based on an extended mathematical input/output calculation model that quantifies the undesirable environmental and social impacts along the entire value chain, this tool links the purchasing and sales data of the Otto Group (including volumes, assortment structure, procurement markets, location, and transport data) with the specific sustainability 'burdens' that the procured and sold goods represent.

The input/output calculation is a concept from economics for which its author, Wassily Leontief, was awarded the Nobel Prize in 1973. With its help, the flow of money between individual business sectors in a national economy—both among themselves as well as through to the end customers—can be described and quantified. Consequently, multi-regional input/output tables allow the value flows caused by Otto Group Purchasing all over the world to be captured in a model. This not only takes the immediate manufacturing of the products procured by the Otto Group into account, but also all the necessary intermediate products, by-products, and primary products.

For this purpose, the monetary purchase volumes as logged by the Otto Group Accounts Department are differentiated by individual products and regions of origin, and linked to the input/output model. On this basis, in a first step, a detailed picture of the direct and induced goods production generated by the Otto Group's business activities all over the world is produced. In a second step, the environmental effects associated with this production are then determined. To do this, the specific environmental burden for each euro of value produced is linked using global environmental statistics to the respective induced production quantity for each individual business sector in each country. Fundamentally, this determines an assessment of the entire Otto Group's ecological scorecard. If we compare the results of the input/output-based methodology with other procedures, we can only see very limited variations. With the input/output methodology, the CO_2 emissions from electricity consumption, transport, and heating, for example, only differ by 4–7% from detailed values raised on the basis of primary physical data.

To make the various adverse effects on the environment comparable, these are finally evaluated in euros. With this innovative monetary evaluation approach we also make environmental matters comparable with 'normal' business topics. Social risks, too, are captured in hard figures through 'estell' with its 'social-risk hours' indicator. To do this, the input/output model is used to determine the working hours for each individual stage of production. The social-risk hours indicator quantifies the respective risks of non-compliance with internationally recognised social standards—in each business sector and country—by means of statistical information. In this case, we do not convert these social-risk hours into terms of financial damages.

Let us take a very simple example that will illustrate the necessary calculation: Suppose the Otto Group was ordering textiles with a purchase value of 1 million euros from China. The physical environmental effects of the respective value chain are now quantified on the basis of statistical data and monetized following the global value interdependencies that the input-output model is reflecting. The aforementioned production, among other things, comes with a footprint of about 1.050t CO_2, 90,000 m^3 of water to be consumed and about 384 kg of particulate matter. These environmental effects are now being evaluated with the regionally specific external costs, which make up a fairly stable data base. Finally, we end with environmental costs of approximately 192,000 € only for water, CO_2 and particulate matter, which are induced by this procurement. 192,000 € or—in other words—19.2% of the purchase volume.

The result of these calculations is a quantitative outline of the principal causes of negative impact along our entire value chain. This helps us to better understand our own actions and to bring suitable measures to bear in the appropriate places in the future. By supplementing this with the stakeholder feedback raised to date, a new characteristic emerges in our established materiality matrix. This is now enhanced by placing the quantitative dimension from 'estell' side by side with the qualitative assessment from our stakeholders. We are not alone in finding this innovative approach convincing: other parties concur, including those stakeholders with whom we have discussed and further developed our project in a dedicated stakeholder dialogue. The German Federal Government was also impressed with our approach and awarded us its CSR Prize in 2014.

The materiality assessment described above is followed by the development of systems and initiatives to effectively address the relevant issues, and efficiently deploy the necessary resources (the 'how'). Examples of this are the Otto Group Social Programme, our Climate Strategy which targets a reduction in the Group's CO_2-emissions by 50% by 2020 (with 2006 as the base year), and our Sustainable Textile Strategy, which includes the requirement to change all own-brand and licenced articles made of cotton to sustainably cultivated cotton. Calculation models based on ecological or life-cycle assessments help us to understand how much these measures reduce environmental damage and social risks. We also know through 'estell' that more than 10% of the total environmental impacts arise with our customers in the utilisation phase. We intend to become even more active here in the next few years through closer communication with our customers, for example, and through appropriate assortment design.

In the third step, the systems developed here are integrated into our company's Corporate Responsibility (CR) strategy and its management processes. Consequently, all Group companies above the voluntary relevance threshold of 100 million euros in annual revenues now have their own CR agenda, with binding target curves that contribute to achieving the Group's objectives. We will be communicating our progress towards these goals internally and externally as part of our Sustainability Communication Programme. Beyond this, these areas of progress are also factored into incentive mechanisms for the members of the Otto Group Executive Board.

4 How Do We Take Opportunities and Positive Contributions into Account, As Well As Any Losses and Risks?

With the approach described above we can effectively reduce the harmful impacts of our business activities, as well as any resulting risks to our business model. But in the end, sustainability must be more than pure risk management. According to the principles of traditional marketing, companies should create direct added value to justify their right to exist. My personal conviction is that this should also apply to the society holistically. This means that the answer to the question of the value of a company would no longer be a sole business issue; it would instead be expressed in terms of macroeconomics and society as a whole. This can be stated in terms of how much added value (economic, environmental, or societal) the company has generated by the end of a year. Where have risks and negative impacts not only been avoided, but also where have positive contributions on balance been made to the common good? Such a view is still the reserve of people who, in the most positive case, are labelled as visionaries—but also quite often as lunatics. I am personally convinced, however, that in the long term there will be no getting around this way of thinking.[1]

[1]This text is based in part on previous publications of the author.

References

Hardoon D et al (2016) An economy for the 1%. Oxfam, Oxford. https://doi.org/10.21201/2016.592643

Steffen W et al (2015) Planetary boundaries: guiding human development on a changing planet. Science 347(6219):1259855

Systain (2017) Die Zukunft erfolgreichen Wirtschaftens liegt in der Lieferkette. http://estell.systain.com. Accessed 29 June 2017

Von Carlowitz HC (1713) SylviculturaOeconomica. Johann Friedrich Braun Verlag, Leipzig

Filippa K's Fashion Manifest: Long Lasting Simplicity

Elin Larsson

For the first time in history, we, the human beings, are now running this ship, planet earth. We are the captains. Scientists say we have entered a new geological era, the Anthropocene—when human activities have a significant global impact on Earth's geology and ecosystems. With that power come a lot of possibilities but also responsibilities. And the solutions of how to take that responsibility vary, as do the reasons for doing so. We at Filippa K are not only doing it because it is the right thing to do but also because we want to run a successful company into the future.

We are a company with a passion for design. Our strategy has always been to make clothes that can last a long time both in style and quality. We offer a style rather than a trend, by creating timeless pieces and wardrobe favourites that can be used season after season.

Our culture is based on strong values, and our products lie at the core of our business. Style, Simplicity, and Quality are three values that characterize everything we do. These values serve as a guide in our everyday efforts to meet our own requirements and those of our stakeholders. At Filippa K, quality means more than just meeting our customers' expectations with regards to our products. Quality is part of everything we do, from how our products are designed, manufactured, and sold, to how we care for our colleagues, customers and everyone else with whom we interact. We strive to be a brand that is worthy of trust. We sense this is something that is appreciated by our customers who are becoming more and more aware, and we are moving towards a society that is much more value-based in its consumption, meaning that people choose brands that have values in line with their own.

We have for many years been working on increasing the number of sustainable fibres, addressing social issues in our supply chain, and decreasing our negative impact in production. Filippa Knutsson, one of the founders of Filippa K, once said,

E. Larsson (✉)
Filippa K, Stockholm, Sweden
e-mail: elin.larsson@filippa-k.se

S. M. Jastram, A.-M. Schneider (eds.), *Sustainable Fashion*, Management for Professionals, https://doi.org/10.1007/978-3-319-74367-7_8

"I set out to build a brand of substance and truth, not dependent on the superficial trends of the fashion industry." That attitude has followed us from the start in 1993. In 2007, we developed our first environmental policy and the following year we opened the first Filippa K Second Hand store.

Furthermore, in 2011 we started an expanded journey with a holistic view, including everything from how we design and develop our products, to how they are produced and how we operate within our own company. We decided on five commitments for the brand that should be achieved by 2030.

By 2030, all materials we use in our products should be more sustainable, not only the main fabric but all trimmings as well, such as lining, buttons, threads, etc. We have developed our own fibre tool where we classify all fibres we use into different categories from 1 to 4, where 1 and 2 are the more sustainable options and 3 and 4 are phase out fibres. This tool helps us to set goals and guides our designers and product developers to make conscious choices during the design phase. We also want to make sure that all styles are recyclable by 2030, meaning that we can take back our clothes and recycle them into new textile fibres to use in our production.

We also want to make sure that our products are produced in a sustainable way. To be able to do that we need to have full traceability and transparency in our value chain. We have not yet set the exact criteria for the what sustainable production means, but we are learning more and more through the projects we do together with our suppliers and plan on doing this during the year 2018.

Knowing that we, the humans, are already using more resources today than the earth can restore we need to start being much more resource efficient within our business. We are part of an industry that leaves significant foot prints and an industry that contributes to a massive over-consumption. One of our biggest contributions to change this is to not produce more than is needed, to have an accurate purchase precision. But being resource efficient also includes everything from how much and what energy we use in our facilities, to what transport methods we choose for our shipments, and to decreasing the usage of paper in our offices.

By 2030 we also want to make sure that all people included in our value chain are treated with respect and have good working conditions, meaning compliance with our Code of Conduct, from the cotton field all the way to the stores.

Being sustainable also means being able to run a healthy and successful business on a long-term basis. To succeed with that goal we need to keep building professional long-term relationships with our partners but also make sure we can deliver a profit so that we can invest in ourselves and have organic growth in our company.

1 Introduction

Filippa K's philosophy is based on the company's three values of Style, Simplicity and Quality. These values apply to all employees, and characterise everything we do. From the design phase, the ambience of our stores, and the way we present the collections, to the way in which we approach our customers, partners and one another.

We work hard to be a brand that our customers and other stakeholders can trust. We strive to be a fair and reliable business partner in a long-term and rewarding relationship based on mutual trust. Our trading agents, suppliers, and other business partners play an important part in realizing this aspiration. All partners of Filippa K are expected to operate in compliance with our Code of Conduct. This means that supplier companies, trading agents, and other partners must ensure that the Code of Conduct is also respected and fulfilled by subcontractors involved in the production processes carried out on behalf of Filippa K.

Our model for corporate responsibility is based on four interrelated and equally important areas: environmental responsibility, social responsibility, financial responsibility and ethical responsibility.

Our environmental responsibility drives us to search for and value partners who strive to do business in an environmentally sustainable way. This partner is characterized by its capacity to continuously reduce its environmental impact. Also, we search for partners that are financially responsible and run their business with a long-term focus. In social responsibility, we value suppliers that take their responsibility as employers seriously and therefore continually work to ensure good labour standards. Filippa K's ethical responsibility will always be a factor in everything we do and is reflected in the way we act in all situations. We think that business relationships with our partners must be based on honesty, openness and respect. We have the ambition for our employees and partners to be able to grow together with us. And therefore, we value suppliers that are open with their processes and challenges so that we can develop together.

Willingness to work in accordance with our social, environmental, and ethical standards in our Code of Conduct is a crucial factor when Filippa K evaluates and selects its suppliers and trading agents. Before entering a business relationship, Filippa K and the concerned partner always discuss correspondence to our Code of Conduct.

Filippa K's goal is to visit each sewing supplier every year to invest in the partner relationship and check compliance with our Code of Conduct. Filippa K is a member company of the Fair Wear Foundation (FWF), an independent, non-profit organisation that works to improve labour conditions for factory employees. FWF conducts audits of Filippa K's suppliers with a focus on social issues since this is their area of expertise (FWF audit clause 1–3, 6, 8–13 in our Code of Conduct). To complement the social audits a third party will conduct audits regarding environmental issues. After an audit, both Filippa K and the specific supplier receives a report that is used as a base for discussion and improvements.

Filippa K's Code of Conduct consists of two levels of requirements. All partners; including trading agents, suppliers, and subcontractors are expected to work in line with all requirements on both levels.

Level one—must have: If a partner violates any of these requirements, then there will be direct and more drastic actions taken to solve the problem. Filippa K reserves the right to take necessary measures to ensure future compliance with the Code of

Conduct. However, we encourage work in order to find a long-term solution. Requirements are:

1. *No child labour*: There shall be no use of child labour. The age for admission to employment shall not be less than the age of completion of compulsory schooling and, in any case, not less than 15 years. (ILO Convention 138) "There shall be no forms of slavery or practices similar to slavery, such as the sale and trafficking of children, debt bondage and serfdom and forced or compulsory labour. [...] Children [in the age of 15–18] shall not perform work which, by its nature or the circumstances in which it is carried out, is likely to harm their health, safety or morals" (ILO 182 1999).
2. *No forced or bonded labour*: There shall be no use of forced, including bonded or prison, labour (ILO convention 29 (1930) and ILO convention 105 (1957)). Nor shall workers be required to lodge "deposits" or their identity papers with their employer.
3. *No severe safety hazards*: The employees and other people in the partner's surroundings shall not be exposed to any direct safety risks, which can lead to long or short-term injuries. Toxic chemicals must be handled in a safe way and all employees must have access to proper protective equipment and working tools. There shall be clear emergency routines, with alarms and exit routes.
4. *No severe environmental pollution*: There shall be no emissions to air, water, or ground that can have a direct severe impact on the surrounding environment, society, or the public health. All partners must treat and manage their emissions in consistency with legal requirements.
5. *No corruption*: Partners should not, directly or indirectly, offer, promise, give, or demand a bribe or other undue advantage to obtain or retain business or other improper advantage. Nor should enterprises be solicited or expected to render a bribe or other undue advantage (OECD Guidelines 6).
6. *Transparency regarding working hours and wages*: All partners must be transparent and willing to show Filippa K documents of employees working hours and wages.
7. *Compliance to Filippa K's chemical restriction list*: Partners are obligated to follow Filippa K Chemical Restriction List (RSL) and continuously phase out chemicals that are added to this list. Chemical containers must be properly labelled and safely stored. A material safety data sheet (MSDS) must be available (in the local language) and the instructions in the MSDS must be followed (ILO 170 1990).

Level two–should have: If a partner violates a level two requirement the partner will be given the opportunity to improve; these improvements must be reported. Thereafter, Filippa K will follow the improvements closely to ensure that this violation is not repeated.

8. *There is no discrimination in employment*: Equality of opportunity and treatment regardless of race, colour, sex, religion, political opinion, nationality, social origin or other distinguishing characteristic shall be provided (ILO

100 1951) and (ILO 111 1958). Discrimination due to gender is a common problem in the textile industry. Filippa K is strongly against sexual harassment and demand that partners treat their workers as equals and give them the same respect, opportunities and salary based on the workers' individual performance.

9. *Freedom of association and the right to collective bargaining are respected*: The right of all workers to form and join trade unions and to bargain collectively should be recognised (ILO 87 1948; ILO 98 1949). Workers' representatives shall not be the subject of discrimination and shall have access to all workplaces necessary to enable them to carry out their representation functions (ILO 135 1971; ILO 143 1971). The representatives must be chosen by the employees and cannot be in a managerial position. Employers shall adopt a positive approach towards the activities of trade unions and an open attitude towards their organisational activities. Workers shall own the decision of whether to join the union.
10. *Living wages are paid*: Wages and benefits paid for a standard working week shall meet at least legal or industry minimum standards and always be sufficient to meet basic needs of workers and their families and to provide some discretionary income. Deductions from wages for disciplinary measures shall not be permitted nor shall any deductions from wages not provided for by national law be permitted without the expressed permission of the worker concerned. All workers shall be provided with written and understandable information about the conditions in respect of wages before they enter employment and of the particulars of their wages for the pay period concerned each time that they are paid.
11. *Hours of work are not excessive*: Hours of work shall comply with applicable laws and industry standards. Working hours should be scheduled to provide ample time for rest, meals, and sleep. In any event, workers shall not be required to work more than 48 h per week on a regular basis and shall be provided with at least 1 day off for every 7-day period. Overtime shall be voluntary, shall not exceed 12 h per week, or 240 h per year, shall not be demanded on a regular basis, and shall always be compensated at a premium rate.
12. *Employment relationship is established*: Obligations to employees under labour or social security laws and regulations arising from the regular employment relationship shall not be avoided through the use of labour-only contracting arrangements, or through apprenticeship schemes where there is no real intent to impart skills or provide regular employment. Younger workers shall be given the opportunity to participate in education and training programmes.
13. *Working conditions are decent*: A safe and hygienic working environment shall be provided, and best occupational health and safety practice shall be promoted, bearing in mind the prevailing knowledge of the industry and of any specific hazards. The working environment should be suitable considering light, noise, climate and ventilation. The working condition shall not lead to any short or long term physical or psychological injuries.
14. *Educate employees*: Partners shall provide adequate education and training to employees in environmental health and safety matters, including the handling of

hazardous materials and the prevention of environmental accidents, as well as more general environmental management areas, such as environmental impact assessment procedures, public relations, and environmental technologies (OECD Guidelines 5.7).

15. *Managing environmental issues*: A management system that addresses environmental issues and drives the business to understand and improve their environmental performance must be established and maintained. This means that suppliers and subcontractors must collect and evaluate information regarding the environmental impacts, and establish measurable objectives and targets for improving their environmental performance (OECD Guidelines 5.1). The management system shall focus on preventing pollution, minimizing waste and emissions, and optimizing water, energy and raw material consumption.[1]

All partners have to sign the receipt of Filippa K's Code of Conduct. They must agree that they comply with all requirements stated in the Code of Conduct. They must also ensure that all subcontractors and suppliers engaged on their behalf comply with the code.

2 Our Vision and Manifest

To succeed with our commitments and try to change the patterns of fashion we realised we needed to make a greater internal change in our view of sustainability.

In 2014 we decided on a new overall vision for the brand; we want to see a fashion where sustainability is the guide to growth, and we developed a manifest telling us how to get there:

If we want fashion to stay relevant and aesthetic, inside and out,
We need to be personal, simple, and long lasting.
Ecosystems are our inspiration.
Planetary boundaries are key, not limitation.
Sustainability is our guide to growth.
We are dedicated to a carefully curated wardrobe built on personal style, circular design and a holistic approach to business.
How we do something is how we do everything.
What we do is long lasting.

During 2017 our vision and purpose of the brand evolved even further to become "To show that simplicity is the purest form of luxury and to inspire a movement of mindful consumption". You can find the manifest in our Sustainability reports. It's based on what we always have been doing, making clothes and accessories that can last for a long time, but it is also built on two new concepts:

[1]All quoted ILO Conventions and OECD guidelines are taken from www.ilo.org and www.oecd.org

The first one, the Curated Wardrobe, is about helping our customers to a smaller, smarter and more selective wardrobe where clothes fit together and are easy to combine and to mix. A wardrobe that you care for and cherish and that can be updated in new ways such as through second hand, leasing, swapping, borrowing etc. And the clothes in that wardrobe never end up in the trash.

The second concept, Circular Fashion, is our internal framework on how to adapt to circular economy, guiding us on how to move away from the linear models that we have today towards circular ones, like nature's eco system.

It includes everything from how we design, develop and produce our clothes to changing our business models and to making sure our clothes really get that long life that they were meant to have and never end up in the trash. How we are starting to transform is described in the following section.

As part of our Circular Fashion concept we are devoted to four R's: reduce, repair, reuse and recycling.

Reduce means that we are determined to create products with minimal negative impact that are part of a circular mind-set. We are constantly questioning what we are doing and challenge ourselves to do things in new ways—like with our front-runner products. Every other year we select a few products to become front-runners for the rest of the collections, serving as guidance on how to transform the rest of the collection and our business. That means we try to make them as sustainable as possible throughout the whole lifecycle and then take the learnings and implement them in the ordinary collections and business. This way we are transforming our business step by step.

Repair is about extending the life length of the clothes within the users. We offer care products to the users in our stores, we talk to them about how to take care of their garments, and we mend broken clothes. We also provide a care guide for our customers, both available as a poster and on our website.

We will most likely always want to update ourselves and change the clothes in our wardrobe before they are worn out. The question is how we can do that without putting the phased-out clothes in the waste bin. We want to make sure that all our products get a chance of a second or third life with another user and that is what reuse is all about, making sure that our used garments get new life by being passed on to a dear friend or sold second hand. Already in 2008, a Filippa K second-hand store opened in Stockholm. Here, customers can sell the clothes they do no longer use and customers can come here to find pieces they missed while available in stores or if their wardrobe favourite has become worn out. The second hand store has a long wish list of clothes that customers are seeking for. We have started to take back clothes through our Collect concept; our customers can hand in Filippa K clothes they do not use any more in any Filippa K store and get a 15% voucher on their next purchase. We make sure to extend the life length of the garment by either selling it through Filippa K second hand or giving it to charity.

Then the question is whether we need to own all the clothes we use? We have started to rent out clothes in selected stores through the Lease concept. The idea is that you and I can renew our wardrobes temporarily without contributing to increased consumption, without getting an overcrowded wardrobe,—and without emptying our pockets! We also support different solutions for collaborative consumption, such as Lånegarderoben (The Closet Library) and Sabina and Friends in Sweden and the fashion library Lena in Holland.

Filippa K clothes are made to last, but sadly they will eventually be worn out. But they still have value—we want our customers to bring old worn-out clothes back so that we can recycle them. As for today we are only able to guarantee recycling of old clothes with our front-runners where we collaborate with a Swedish start-up company called Re:Newcell that is developing a textile to textile recycling method for cellulose fabrics and an Italian company that can recycle our front-runners made of wool. Beyond that we have started to reuse cutting waste. Several of our suppliers are sending their cutting waste to our warehouse DHL, who gathers the waste and then ship it to a fabric manufacturer in Italy for reuse.

Closing material loops in order to minimize the amount of textiles that ends up in landfills or gets incinerated is a big challenge that requires new infrastructural solutions. We cannot do it on our own, so we cooperate with other actors to improve the possibilities of recycling textile fibres in the future. Some examples of actors with which we cooperate are re:newcell, the Swedish government bodies Kemi and Naturvårdsverket (Swedish Environmental Protection Agency), and the Nordic Waste Group, Swerea, and Mistra Future Fashion.

3 Innovation and Collaboration

There are many challenges in being part of the textile industry, an industry with a long and complex supply chain and one that leaves significant environmental and social footprints. We constantly challenge ourselves to reduce our negative impacts through the choice of our materials, our practices in production, and our partners in the value chain. But perhaps most importantly, our aim is to not produce more than is needed and to not contribute to overconsumption. For us, that means increasing our purchase precision, and offering long-lasting products of quality, style, and simplicity, as well as enabling our products to a second life, for example through our second hand store. And this is where our vision and manifest help and guide us on what to do and where to focus.

We see ourselves as one part of a larger ecosystem, where each part is equally important and each part has its unique knowledge to contribute. Our ecosystem consists of our employees, our owners, our suppliers and partners, NGOs and experts, government bodies, and last but not least, our customers. It is crucial to interact and have a dialogue with all of these stakeholders. For instance, we have been working with Fair Wear Foundation since 2008 to secure good labour standards in our supply chain, as well as better compliance with our Code of Conduct. Fair Wear Foundation is a multi-stakeholder initiative that supports us to strive for secure good labour standards at the sewing factories that are located in Europe 71% and the rest of the world, mainly in China, but also some small production sites in India, Peru, and Vietnam. We often use the FWF to perform factory audits, and we also utilise their expertise when carrying out reports and action plans if needed. Also, we use FWF representatives to perform the audits, to ensure that these are professionally executed, and to have the supplier understand that we take the auditing seriously. The audits help our suppliers to develop and improve their social compliance standards. But most importantly, we work proactively with these issues with help from FWF to provide valuable trainings to raise awareness among the factory workers. We have always strived to build long-term relationships with our partners and to keep the number of suppliers down. That makes the quality of our relationships, as well as the products, on a higher level.

The area of gamification is a growing source of inspiration in attempts to make fashion more sustainable. Gamification is the application of game principles in non-game contexts, and is most commonly used in attempts to improve learning, organizational productivity, evaluation and user engagement. Applying gamified services in factories producing clothes, like in the case of QuizRR (Quizrr 2017), which we are a part of, creates wider knowledge amongst workers and hence potentially improves their working conditions. QuizRR is a digital learning tool that is used on a tablet that the factory receives. It contains pedagogic introductions movies to each theme and that is followed up with questions where you can choose from three different answers. When answering correctly, you are rewarded with stars, if incorrect, you can try again. You do the quiz anonymously but register the group within the company to which you belong. Increased knowledge about rights and responsibilities are widely needed among factory workers in developing markets, and factories often lack functioning systems for workplace rights and safety information. Using QuizRR makes it possible for the factory to raise awareness amongst the workers (both management and floor workers) without needing to shut down the factory for education hours, because the employees can use the tablet when suitable and you can have a wider impact on and increasing number of engaged workers.

Gamified services can also help suppliers and brands gain information about knowledge gaps and other "weak spots" in their organization. We have also tried out gamification within our own company through an online training created by an organisation called Sustainable Fashion Academy where employees take part of a sustainability course. The purpose of the course is to provide brand and retail professionals the foundational knowledge and inspiration they need to engage with sustainability challenges and opportunities within their companies, and to become more confident organizational ambassadors for sustainability.

Key factors for a sustainable fashion future:

1. *Knowledge creates engagement*: Sustainability can be an overwhelming subject and sometimes almost frightening: where to begin, a lot of difficult words, so many areas of concern, etc. In order to create change, no matter in what part of the eco system, raising awareness is crucial. I believe this quote from Sean Paul Sartre describes it very well, "Once we know and are aware, we are responsible for our action and our inaction. We can do something about it or ignore it. Either way, we are still responsible." (AZQuotes 2017)

2. *Fully integrated vision and clear direction*: Once you are aware you need a clear vision and direction to know what to focus on and to make sure the company knows where to go. Since the start, we have decided not to have a separate sustainability department, but instead make sure sustainability is integrated and a part of everyone's work and processes. By doing so sustainability becomes a natural part of the core business and decisions.
3. *Clear strategy and concept*: For everyone to be on board and to contribute to the vision, you need a clear and simple strategy to which everyone can relate. I like to compare the strategy with a Christmas tree. Before you have the strategy it is like you have all those beautiful Christmas ornaments, but nowhere to hang them and therefore you do not use them, but that changes once you have the Christmas tree in place and everyone can start decorating the tree. That is why it is so important to define a clear strategy.
4. *Be bold enough to try new things*: Breaking new grounds can be tough and you need to have some guts to challenge yourselves and dare to try out new things. I believe that the employees at Filippa K show a lot of that; we are constantly challenging ourselves and trying out new ideas and solutions. It also takes a lot of patience and conviction to make change happen, at least to get the majority on board. The road can be bumpy and winding sometimes, but that is part of the transformation phase. We also found out that it is easier to try out new ideas and solutions on a smaller scale and as separate projects from the beginning. That way we do not risk too much when trying out new things. It is important though that the ideas are scalable, because we always have the intention of growth and of developing the idea to become a contributing part of our business, no matter if it concerns product development or new services to our customers. For us it is crucial to participate in different international forums, to visit interesting seminars and to look at what other industries are doing in order to get inspiration for new solutions. Without that our business development would stagnate.
5. *Be humble enough to adjust along the way*: When developing completely new ideas and changes it is difficult to know how it will work in reality. How customers will perceive it, how well it will integrate with the rest of the business, unforeseen logistical challenges, etc. That is why you need to allow yourselves to learn as you go along and make adjustments along the way or remove it if it is not successful. An important ingredient is also patience and endurance; to let the new initiatives take the time they need, it is not common to see a breakthrough right away. One of the most important objectives of our innovations is that it is a learning process; let the lessons and insights spread throughout the business and create a bigger change than just within the specific project. I met a Swedish Buddhist monk once who told me a story about a little girl. The 7-year old girl walked all alone on a beach. There had been a dreadful storm the night before and thousands of starfish had been washed up on the beach. The little girl picked them up one by one to throw them back into the sea when an old man walks by. He says to her, "Stop that, there is no use, you cannot save them all anyway". The little girl bends down, picks up another starfish and throws it into the sea and says, "Ask that one what he thinks". I like that story a lot, because it says a lot about the

importance of the small actions and how they can spread and create a greater change in the end.

Time is also essential to succeed but also one of the biggest challenges in most companies. You need time for reflection and inspiration. You need time for learning what needs to be fixed. You need time for creating those innovative ideas, and you need time to implement them and to create the desirable change.

We have also learned the importance of making innovation and sustainability an integrated part of the company and everyone's daily work, to not have a separate sustainability department. It is by engaging people and setting up common goals that you can create a real change. And that means to work top down but also bottom up, to succeed. You need the management as well as the experts within each part of your business to be engaged and motivated.

4 Time for a Change

We strive to run a long-term sustainable business within the planetary boundaries. To ensure long-term sustainable success we must have a holistic view of our business, understand how all parts interact and make sure our value chains are sustainable long-term. To include and collaborate with all stakeholders in our ecosystem and to get the customers on board is crucial. Because we cannot create this change by ourselves. Our last initiative was to invite our customers to become test pilots—to try out our different products and services to be able to give us feedback on how to improve and develop ourselves further. The response we got is amazing! They want to act and to have an influence on the way forward and we are so thankful for that—together we can move mountains.

It is a big system change that is needed and here we can learn from nature. Nature is already working in circles and has been doing so for at least 3.5 billion years and had plenty of time to fine tune and adjust. That is why biomimicry is so inspiring; nature shows us that the answers are all around us, with species willing to share their solutions.

Together with our owners, partners and the rest of the industry we want to challenge the conventional fashion industry and show the way to a more sustainable, more profitable and more innovative fashion, a fashion built on circular models.

We are discovering new grounds, grounds without any maps because it is all so new. Our journey is built on conviction, collaboration, and creativity, but also courage, tenacity, and hard work. And that is something everyone at Filippa K and our owners have shown. We are looking forward to expanding the limits of what we thought was possible, together with our partners, the rest of the industry and our customers.

I would like to finish with a divination for the future of fashion: A day in the life in 2020.

It's Friday. You pick out clothes for the day from your minimal wardrobe that, besides a good base of sustainable basics, consists of rented or borrowed clothes; stand-out pieces that are only there for a limited time. During your lunch break, you go to a Filippa K hub, not only to have organic coffee but also to develop a design for a one-of-a-kind disposable top with the help of a designer. You print it on your 3D printer and decide not get it delivered to your house by Uber but instead to put it in your climate positive bag made of carbon dioxide from the air. Once home, you see that someone has put in a request to rent one of the pieces in your virtual closet. You accept the request and put it in the safety box for pickup outside your door that is opened by fingerprint. You discover a stain on your jacket and take out your miniature spot-clean steamer to clean and refresh the jacket on the go, or perhaps put it in your rented non-water washing machine that cleans with the help of carbon dioxide.

It's a good party, you dance the night away and sweat a lot without your disposable top breaking or falling off. Back home, you put it in the recycling bin with yesterday's newspaper.

Since you're going on a business trip on Sunday you take the opportunity to get a proper workout on Saturday. You go for a long run and your sportswear warns when you need added energy to endure. You take a bite out of the sleeve of the long-sleeved running top, it tastes like banana. Sunday comes and you travel to Portugal without luggage. At Faro airport, you pick up a bag with clothes to use during the coming week, rented online during the flight.

To look at where we are right now and how to build a more sustainable wardrobe, I would recommend buying versatile pieces that you find valuable, that can become the backbone of your wardrobe. Make sure to care for those clothes so they can have the long life that they were meant to have. You can update that wardrobe in new ways such as through second hand, rental, or swapping with your friends.

We are still in the beginning of our journey and we have a long way to go, but we are highly motivated and determined to succeed. Our industry needs to change. Either you choose to ignore that or you become part of the transformation, and that is what we have chosen.

References

AZQuotes (2017) http://www.azquotes.com/quote/867303. Accessed 15 June 2017

ILO 29 (1930) CO29 – Forced Labour Convention, 1930. http://www.ilo.org/dyn/normlex/en/f?p=NORMLEXPUB:12100:0::NO::P12100_ILO_CODE:C029. Accessed 19 June 2017

ILO 87 (1948) C087 – Freedom of Association and Protection of the Right to Organise Convention, 1948 (No. 87). http://www.ilo.org/dyn/normlex/en/f?p=NORMLEXPUB:12100:0::NO::P12100_INSTRUMENT_ID:312232. Accessed 15 June 2017

ILO 98 (1949) C098 – Right to Organise and Collective Bargaining Convention, 1949 (No. 98). http://www.ilo.org/dyn/normlex/en/f?p=NORMLEXPUB:12100:0::NO::P12100_ILO_CODE:C098. Accessed 15 June 2017

ILO 100 (1951) C100 – Equal Remuneration Convention, 1951 (No. 100). http://www.ilo.org/dyn/normlex/en/f?p=NORMLEXPUB:12100:0::NO::P12100_ILO_CODE:C100. Accessed 15 June 2017
ILO 105 (1957) C105 – Abolition of Forced Labour Convention. http://www.ilo.org/dyn/normlex/en/f?p=NORMLEXPUB:12100:0::NO::P12100_ILO_CODE:C105. Accessed 19 June 2017
ILO 111 (1958) C111 – Discrimination (Employment and Occupation) Convention, 1958 (No. 111). http://www.ilo.org/dyn/normlex/en/f?p=NORMLEXPUB:12100:0::NO::P12100_ILO_CODE:C111. Accessed 15 June 2017
ILO 135 (1971) C135 – Workers' Representatives Convention, 1971 (No. 135). http://www.ilo.org/dyn/normlex/en/f?p=NORMLEXPUB:12100:0::NO::P12100_ILO_CODE:C135. Accessed 15 June 2017
ILO 143 (1971) R143 – Workers' Representatives Recommendation, 1971 (No. 143). http://www.ilo.org/dyn/normlex/en/f?p=1000:12100:::NO:12100:P12100_INSTRUMENT_ID:312481. Accessed 15 June 2017
ILO 170 (1990) C170 – Chemicals Convention, 1990 (No. 170). http://www.ilo.org/dyn/normlex/en/f?p=NORMLEXPUB:12100:0::NO::P12100_ILO_CODE:C170. Accessed 15 June 2017
ILO 182 (1999) C182 – Worst Forms of Child Labour Convention, 1999 (No. 182). http://www.ilo.org/dyn/normlex/en/f?p=NORMLEXPUB:12100:0::NO::P12100_ILO_CODE:C182. Accessed 15 June 2017
Quizrr (2017) https://www.quizrr.se. Accessed 15 June 2017

Sustainability Strategies in Luxury Fashion: Company Disclosure on Human Rights

Isabel Ebert

1 Introduction

At first sight, luxury fashion might appear as the ideal precedence case for a sector that has the potential to become a pioneer in sustainable business. Luxury fashion includes fashion products that are in contrast with necessity goods, typically costlier, and predominantly purchased by customers with an elevated disposable income or greater accumulated wealth than the average population.

There is no clear-cut definition from which price range a piece of fashion can be considered luxury as the concept closely interrelates with individual perception of what is a luxury lifestyle. Nevertheless, some key features tend to characterize luxury fashion objects, which in turn make the luxury sector, in theory, very suitable to step up for sustainability.

The presumed high quality of luxury fashion implies a longer use of the fashion piece, which spreads the environmental and social costs for the production of the item on a longer time span (Joy et al. 2012). Hence, in contrast with a mainstream brand that targets a customer wanting to wear its products only several times (fast fashion model), luxury brands want to guarantee longevity of their products over many years. With regard to human rights and working conditions in fashion production in particular, the elevated pricing policy of luxury fashion brands a priori permits more scope for integrating additional costs for sustainable manufacturing into the production process. For example, the additional costs for in-depth working conditions assessments may be transferred onto the final product price without making a significant difference for the actual purchaser of the product due to her or his high disposable income. In contrast to mainstream brands, the lowest competitive price is not a key argument for selling luxury products. Furthermore, one

I. Ebert (✉)
Institute for Business Ethics, University of St. Gallen, St. Gallen, Switzerland
e-mail: isabel.ebert@unisg.ch

S. M. Jastram, A.-M. Schneider (eds.), *Sustainable Fashion*, Management for Professionals, https://doi.org/10.1007/978-3-319-74367-7_9

could argue that structural labour issues in apparel manufacturing in the Global South affect luxury brands less, as production sites are traditionally located in the Western Hemisphere (e.g. Italy, France, Spain, Portugal). Besides, brand reputation in luxury fashion is highly relevant for sales and the vulnerability for reputation loss is significant—to a certain extent comparable to the volatility in the diamond industry (compare engagement of the diamond industry in the Kimberley process as analysed by Kantz 2007). Products need to embody a high degree of exclusivity in design, together with impeccable quality standards. Hence, if a luxury fashion company fails to ensure socially sustainable production conditions, how could it be guaranteed that exclusivity in design and quality are assured?

Having a closer look, the factual evidence on luxury fashion's performance on sustainability, in particular human rights, draws a different picture. Given that the vast majority of luxury fashion companies are outsourcing their production similar to common high street fashion companies, luxury companies become more vulnerable to alleged environmental and social rights abuses. The luxury sector is not immune to social issues that prevail at the most vulnerable spots in global fashion supply chains (e.g. Clean Clothes Campaign 2014) and there are even calls to reform trademark law so that luxury companies meet their ethical responsibilities (Sun 2013). This paper aims to convey an overview of the degree of transparency to which leading luxury companies disclose information on their sustainability policies and management approaches, with a specific focus on human rights, and social and labour standards on the manufacturing stage. Transparency is crucial for pushing sustainability in the industry forward as it allows investors and consumers alike to make their investment or purchase decisions respectively on solid grounds. It also allows competitors and allies in the business to benchmark their performance within the industry and learn from each other's practices to push for progressive company performance and industry standards.

2 Structural Factors in Luxury Fashion with Regard to Sustainability

The marketing approach of luxury brands refers to key factors of quality, exclusive availability of the product, the precise and expert craftsmanship process, and product innovation, correlating with and to a certain extent also justifying the elevated price range of the purchased product:

> The marketing strategies of the luxury industry have been the door for their consumers to the wonderland of luxury spending and consumption. These strategies are meticulously designed to highlight high quality, exclusivity, craftsmanship, precision, and innovation that luxury goods can deliver to their customers (Sun 2013: 554).

Trademarks appear as a crucial factor. Trademarks stand for the authenticity and hence exclusivity of the luxury fashion item, which is key for the consumer. Adding to this, the tracking of signature items (e.g. individual numbers for handbags) to ensure quality bears the potential to enable more effective safeguarding of social sustainability indicators. Even with the shift of production to low labour cost regions such as China, Romania, or Saudi Arabia, it remains important for luxury companies to capitalize on Italian or French labour jurisdictions in a way that permits maintenance of the "Made in Italy" or "Made in France" label (Likoksy 2008: 294). With the increased outsourcing of manufacturing of luxury fashion in the last 20–30 years, seamstress guilds in the Western Hemisphere with union-like rights have been replaced by (mostly) women working under debatable conditions in Vietnam or Cambodia, and other countries (ibid: 298). Civil society reports, such as a recent one by Clean Clothes Campaign (2014), raise harsh criticism against luxury brands. The report by Clean Clothes Campaign alleges that workers are often paid wages below any established standard and are required to work very long hours in factories that fail to meet basic health and safety requirements, with labour conditions in some more extreme cases equalling conditions of modern slavery. Adding to this, in order to maintain the local trademark labels, such as the "Made in Italy" label, migrant workers or second or third generation migrant workers inside Europe become essential. For example, there are a myriad of Chinese migrant workers in Italy (ibid: 19). According to Clean Clothes Campaign, since a first wave of immigration during the period between the world wars and in a second wave in the 1980s and 1990s, the Chinese diaspora has become an important element of the competitiveness of the Italian fashion sector as a whole. A network of Chinese family businesses is supplying Italian manufacturers in a flexible and lowcost manner, mostly working under Chinese rather than European labour conditions. Luxury companies cannot be excluded from these market dynamics (ibid: 20). Ceccagno and Rastrelli (2008: 72f.) cite that for instance, in the historic Prato in Tuscany, Chinese-run companies represent about 80% of all active enterprises. Clean Clothes Campaign (2014: 20) stresses that the supply chains and business relationships between Chinese businesses and their Italian clients are characterised by a high degree of fragmentation and complexity, excessive subcontracting, and labour conditions that can hardly be monitored or controlled. Also, the organization claims that Pakistani suppliers are flowing into the country, establishing similar structures as the Chinese, for instance in the Campania region (ibid). According to Clean Clothes Campaign, French luxury brands equally source from these foreign-run subcontracting factories in Italy (ibid). To ensure compliance with legal provisions on labour conditions, law enforcement agencies and labour inspectors are carrying out various operations (ibid: 22).

These allegations, such as the ones described in the already mentioned report by Clean Clothes Campaign, put pressure on luxury fashion companies to act. At the same time, the companies are also pushed to becoming even more and more profitable. This relates to the increasing extension of businesses into emerging markets (Likoksy 2008). Similar to other major global companies, such as Starbucks or Warner Brothers, luxury fashion brands aim at tailoring their products to the local context for emerging markets consumers beyond the products the companies are offering to customers in their more traditional European or North American market since decades. Hence, besides the outsourcing dynamics in production, another transformation is taking place: the attempt to adapt products to recently arising markets. This expansion puts additional financing pressure on businesses. As a result of the market pressure to react fast to these new demands, luxury fashion companies work increasingly with outsourcing professionals, banks, multinational law firms, and accountancies (Likoksy 2008: 299). Additionally, fashion companies are more and more publicly-traded rather than privately-held. Hence, the character of the luxury business model itself is changing (ibid 299). The profit cycle expectation shifted fundamentally as fashion houses decided to go public. Before, profitability of a luxury fashion business had traditionally been measured over years if not decades. Even a collection that did not sell well would have been accepted as an exception by a fashion house, because success was measured by more long term criteria than it is now. In other words, attached to decisions to be a publicly traded company or to be managed by a non-luxury executive triggered a shift in accountability rules. As in other sectors facing similar restructuring by financial institutions, the aim of investors has been to turn a quick profit from fashion. Decision-making changed to a short-term basis model, nowadays to a three-to-four month cycle of profitability (Likoksy: 299).

Structural factors are essential for a socially sustainable conduct of a company as these influence the corporate culture and processes within the company. For instance, with increased profit cycle expectations, luxury fashion companies need to cut down on financial resources deployed on production and hence among others also labour costs. Thus, decent labour conditions appear less of a priority in the light of financial pressure.

3 Increasing Demand for Transparency

Along with reports about the challenges of luxury companies to respect human rights and labour rights along their value chains, various initiatives and international frameworks exist to provide some broader orientation for addressing challenges to respect human rights. The responsibilities of business to respect human rights are defined in the UN Guiding Principles on Business and Human Rights (United Nations Human Rights Council 2011), along with its major proponent John Ruggie's introduction on the responsibility of business to respect Human Rights (Ruggie 2013). Moreover one could name the International Labour Organization's Declaration on Fundamental Principles and Rights at Work (ILO 1998), the principles of the UN Global Compact (United Nations 2016) and the Organisation for Economic Co-operation and Development's (OECD) Guidelines on Multinational Enterprises (OECD 1976/2016) serve as further key reference points for human rights obligations of companies and labour standards. Sector specific initiatives for the textile industry include for example the Fair Labour Association (FLA), Fair Wear Foundation, the German Partnership for Sustainable Textiles and the Dutch Textile Covenant. For retail in general some key initiatives are the Ethical Trading Initiative (Ethical Trading Initiative and Hult International Business School 2016) or KnowTheChain (KTC) with a focus on forced labour in global supply chains (KnowTheChain 2016a, b). In addition, investors, consumers, and civil society increasingly demand transparency from luxury brands in terms of social conduct and monitor their performance (Winston 2016). In 1991, the Down Jones Sustainability Index emerged as the first global sustainability benchmark and prevails as the key reference point for investors and companies in terms of sustainable investment (RobecoSAM 2016). In terms of benchmarks on human rights, there is the Corporate Human Rights Benchmark, which includes as set of luxury fashion companies (Corporate Human Rights Benchmark 2016).

Also academics speak out about the need for luxury brands to weigh in on promoting social sustainability at the core of their value chains—along with durability, quality and style (Joy et al. 2012).

4 Information Disclosure by Luxury Fashion Companies

In light of these structural challenges for establishing sustainable strategies in luxury fashion, how transparent are luxury fashion companies about their sustainability approach, especially with regard to human rights at work? How much information do they disclose? Based on a sample selection of luxury fashion companies, this study aims to convey further insights by a comparative approach of analysing disclosed information by the companies. The following ten luxury fashion companies are according to a study by EY the leading ones in terms of market capitalisation and will hence be used for this analysis (EY 2015): Burberry, Coach, Hermès, Hugo Boss, Kering, LVMH Moet Hennessy Louis Vuitton, Michael Kors, Prada, Ralph Lauren and Salvatore Ferragamo.

The analysis serves as a snapshot in time (Summer 2016) and relies on company information available in the public domain, e.g. annual reports, websites, and is gathered to indicate how companies address human rights on a policy and management level inside and outside the corporations. Different corporate structures and terms as well as the companies' individual operating contexts make an all-encompassing comparison a challenging task. Not every component of one company policy would equal a component of another company's policy, thus comparisons are only possible to a limited extent. Moreover, as the focus of this study lies on transparency in a company's disclosure of information on social standards, labour standards, and human rights, the measurement of actual social sustainability performance is not part of this paper. This would require an in-depth analysis of each company combined with its particular local structure at each production site and its context. Rather, the focus of this paper is an assessment of the scope and the degree of detail of disclosed information on human rights policies and practices by companies.

Therefore, this paper analyses whether a company discloses comprehensive information in line with the UN Guiding Principles on Business and Human Rights (United Nations Human Rights Council 2011). This paper depicts which degree of commitment the company's information displays on social standards, labour standards and human rights as a first step, secondly whether it has a human rights policy

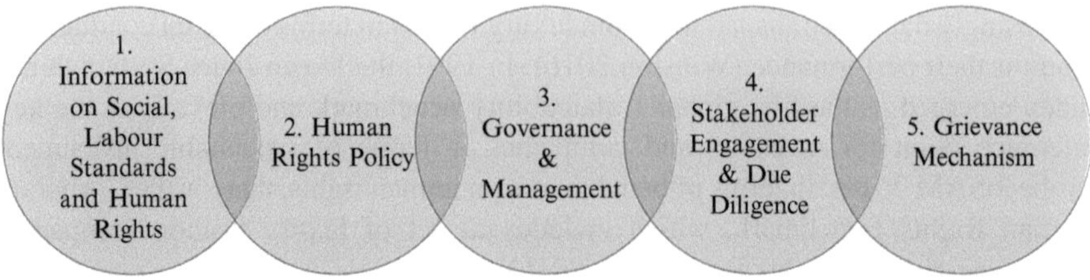

Fig. 1 Dimensions of corporate information disclosure on social sustainability (own illustration)

Table 1 Comparison of company disclosure by dimension for each company

Company name	1. Information on social, labour standards and human rights	2. Human rights policy	3. Governance and management	4. Stakeholder engagement and due diligence	5. Grievance mechanism
Burberry	✓	✓	✓	✓	✓
Coach	✓	✗	✗	✓	✗
Hermès	✗	✗	✗	✗	✗
Hugo Boss	✓	✓	✓	✓	✓
Kering	✓	✗	✗	✓	✗
LVMH	✓	✓	✗	✗	✗
Michael Kors	✗	✗	✗	✗	✗
Prada	✓	✗	✗	✗	✗
Ralph Lauren	✓	✓	✓	✓	✓
Salvatore Ferragamo	✗	✗	✗	✗	✗
Total	7	4	3	5	3

in place that is publicly communicated, thirdly whether it conveys information on the governance and management of the policy's contents, fourthly how a company indicates to engage with stakeholder to carry out due diligence on Human Rights concerns and fifthly whether there is a grievance mechanism in place and whether information on it is publicly available as this paper's focus lies on disclosure (compare Fig. 1). The fashion companies included in this analysis will be clustered into three groups: laggards, catching-ups and front-runners. Laggards address less than two dimensions of corporate disclosure on social sustainability whereas catching-up companies fulfil at least two and up to four out of the five criteria (compare Table 1). The frontrunners disclose information on all five dimensions (Table 1).

5 Empirical Results

Burberry presents an entire media portal on corporate responsibility (Burberry 2016), publicly available on its website. Within this portal, it lays out its Human Rights Policy (ibid), which includes a comprehensive section on how human rights are managed within the company as well as how workers can file their complaints through available grievance mechanisms. Burberry indicates to uphold the rights for all workers in its supply chain to join a trade union through the Ethical Trading Programme (compare ibid). Burberry states that the company requires factories to provide a grievance mechanism, which is communicated to, understood by and applied fairly to their employees—verified by the activities of the Ethical Trading Programme (compare ibid).

Coach's website has no general human rights policy publicly available, and does not disclose precise information on social standards, labour standards, or human rights by the company, although some information can be found on engagement with business partners and contractors. The Project Just, a fashion consumer awareness project, points out that all of Coach's business partners are subject to a contractual requirement to comply with the company's social compliance guidelines (Project Just 2016). However, these guidelines are not publicly available—the brand solely shares its supplier selection guidelines as part of its supply chain stewardship programme (Coach 2016: 25f). Coach indicates that the social compliance guidelines are an enhanced version of the supplier selection guidelines (Project Just 2016). Coach claims to not knowingly work with suppliers who fail to pay their employees a local minimum wage and is partnering with *Better Work Vietnam* to educate, improve, and maintain positive working conditions in some of its factories (ibid) (Table 1).

Hermès discloses hardly any information on human rights within its annual report or in any other sources available in the public domain (Hermès 2016).

Hugo Boss discloses factory assessments in cooperation with the Fair Labour Association which is a significant transparency commitment, and has a Sustainability Committee in place. The company is also a member of the Bangladesh ACCORD on Building and Fire Safety, as well as of the German Partnership for Sustainable Textiles and the Global Social Compliance Programme (GSCP). The company furthermore engages with workers over an ombudsman system. Hugo Boss depicts comprehensive information under the KTC disclosure platform (KnowTheChain 2016b). The company claims to engage with workers in its supply chain on labour-related issues during social compliance audits, in which worker interviews are an important part and which focus on general working conditions as well as on the verification of work-related documents (ibid). Hugo Boss points out that auditors hand out contact cards with the e-mail address of Hugo Boss to the employees so that workers can file complaints independent of the audit (ibid). On a more advanced level of grievance mechanisms, Hugo Boss participates in the FLA, which enables workers to use the anonymous complaint procedure provided by the FLA. This also obliges Hugo Boss to investigate, follow-up, and remedy the cases filed under the FLA complaint procedure. Hugo Boss also points out that it established "a global ombudsman system as a supplementary reporting channel. Employees and third parties such as suppliers or customers can confidentially notify an ombudsman, if there are indications of fraud, infringements of antitrust law or breaches of compliance guidelines. It is also possible to remain anonymous if desired" (KnowTheChain 2016b).

The Kering Group presents an informative sustainability report (Kering 2016), which is at the same time not very precise on Human Rights, governance and management of the latter, as well as grievance mechanisms. Regarding stakeholder engagement, the group highlights that a training programme on ethics was set up in 2013, together with a code for all of the group's employees. It is available in nine languages and lays out the ethical ground rules at Kering along with case studies and exemplary ethical dilemmas (KnowTheChain 2016b). The training programme was introduced in February 2014, with the 2014 edition focussing on corruption, fraud, conflict of interest, and confidentiality of information on social media. Respect for human rights, diversity, corruption, and protection of the environment were emphasized in 2015, and responsible sourcing of raw materials, traceability, and compliance with business confidentiality in the subsequent year (ibid). Remarkably, the Kering Group repeatedly references PUMA as a leading and innovative company within the conglomerate when it comes to social issues. At the same time, this does not excuse other brands within Kering's portfolio from not disclosing information on human rights conduct.

LVMH Moet Hennessy Louis Vuitton lays out multiple activities in its Social Responsibility Report, although the only precise information contained on social issues might trace back to its membership of the UN Global Compact and to its statement on human rights as part of its code of conduct (LVMH 2016). LVMH disclosed information to the UNGP reporting database (Shift and Mazars 2016). With a similar corporate set-up as Kering, LVMH is a big conglomerate of brands.

Michael Kors does not disclose much information on its website, nor in its annual report, on social standards, labour rights, and/or human rights (Kors 2016).

Prada discloses only a glimpse of information on social issues, and regularly points to its Investors Report and Code of Ethics, analogous to when contacted by KTC (Prada 2016). The term "human rights" does not appear in the company's code of ethics; instead, the company refers to respect towards the "human person".

Ralph Lauren conveys insightful information on its Global Human Rights Compliance Department and indicates to carry out capacity building and factory efficiency trainings (Lauren 2016). The company promises to extensively review all first tier suppliers on a mandatory basis, which involves an assessment of risks and recommendations for corrective actions processes (KnowTheChain 2016b). Additionally, Ralph Lauren has review processes in place for foreign migrant workers in higher-risk environments, for instance in Jordan and in India (ibid). Also, the company indicates that it is developing traceability policies with respect to certain product commodities and requests suppliers to put in place grievance mechanisms for workers (ibid). These policies are said to include procedures for risk assessments, as well as for reviews of employment practices and workers' working conditions (ibid).

Salvatore Ferragamo conveys hardly any information on social issues, except some broad lines in the company's financial report in the public domain (Ferragamo 2016).

6 Comparative Summary: Transparency on Sustainability

A general strategic tendency across all fashion companies is hard to detect although this paper can identify different sub-clusters with regard to information disclosure by the luxury fashion companies in the study sample. While disclosure on human rights and labour issues can give some indication of a company's commitment to an engagement on human rights, it is important to bear in mind that the degree of information disclosure might not necessarily equal a company's actual performance with regard to respecting human rights.

The least progressive companies in this analysis, the "*laggards*" encompass Hermès, Michael Kors, Prada, and Salvatore Ferragamo. These companies failed to disclose precise information on their policies and practices to respect human rights, or social or labour standards.

On the "*catching-up*" horizon are Coach, Kering and LVMH as the companies are taking action to manage human rights issues with an increasing degree of importance attributed to the issue. Coach only shows a mixed record of information disclosure, highlighting material such as the social compliance guideline, which are not publicly available though at the same time stating it is an enhanced version of the publicly available supplier selection guidelines. Such an approach can be called biased. At the same time, the public availability of the supplier selection guidelines puts pressure on the suppliers and makes them prepare their production sites in line with the guidelines. Kering has engaged its employees recently in training programmes that could develop into a crucial mechanism for stakeholder engagement to carry out due diligence. At the current stage though, the precise action taken by Kering to ensure the respect for human rights lacks disclosure. There is potential that less progressive brands within the group could learn from more advanced brands with a solid record of respecting of human rights. LVMH appears to engage in some activities on the human rights agenda as part of its UN Global Compact signatory status, although the level of information disclosure remains low, apart from its disclosure on the UNGP reporting database. Hence LVMH needs to be placed on the lower end of the "catching up" continuum.

Burberry, Hugo Boss, and Ralph Lauren stand out as the "*front-runners*" within the analysed group of luxury companies. Each of these fashion companies discloses considerable amounts of information on social, labour, and human rights standards, lays out details on their human rights policy and how human rights are managed within the company, as well as how the company engages with stakeholders to carry out due diligence on human rights. Each of the companies has some sort of grievance mechanism in place, although the quality criteria among the grievance procedures differ even in this advanced cluster. It appears that Burberry and Ralph Lauren convey more information on the nature of their grievance mechanisms, indicating for example corrective action plans, than Hugo Boss.

In this analysis we can see that the information on social, labour standards, and human rights was conveyed by seven out of ten companies. With a more precise degree of information demanded on human rights policy, the number of companies disclosing on this dimension decreased. Five out of ten companies disclosed information under dimension of Stakeholder Engagement and Due Diligence. Four out of ten companies had a human rights policy commitment. The dimension of management and governance was equally the least fulfilled together with grievance mechanisms (three out of ten companies). To a certain degree this might indicate that a strategic approach for broader integration of human rights governance into the overall company strategy and handling of complaints is not yet incorporated in many company strategies.

Social sustainability is relevant—not only for a company's reputation and investor interests but also for internal strategic processes. The luxury fashion sector has a distinct responsibility—and opportunity—to emerge as a pioneer in the textile sector to respect human rights. Due to its elevated pricing policy and high quality positioning, luxury brands could embody the case of differentiation from mainstream brands by an impeccable human rights performance. There is immense unused potential in the luxury fashion sector that can be unlocked to foster social sustainability. To achieve this, luxury brands need to comprehensively integrate social sustainability in a comprehensive approach into their corporate strategy. This paper analysed the external disclosure of fashion companies on social sustainability, thereby touching on publicly available human rights commitments and other documents that target policies and practices. Of further relevance would be an in-depth look into the actual emerging of corporate strategy on social sustainability by luxury brands. Such insight would make it possible to compare which strategic approaches are applied to address social sustainability and in a next step, to which effects these strategic approaches lead to performance of social sustainability "on the ground". This paper shows that some of the analysed fashion companies are stepping up to become front-runners, at least in their policies and practices formulated "on paper". At the same time, it takes more to make luxury fashion an industry that embodies the slogan "fashion is made by humans for humans" (from German by Christiane Arp in Vogue September 2016).

References

Burberry (2016) Corporate responsibility. Human Rights. http://www.burberryplc.com/corporate_responsibility, http://www.burberryplc.com/documents/corporate_responsibility/human-rights-policy.pdf. Accessed July 2016

Ceccagno A, Rastrelli R (2008) Ombre cinesi? Dinamichemigratoriedella diaspora cinese in Italia. Carocci, Rome

Clean Clothes Campaign (2014) Can you earn a living wage in fashion in Italy. https://cleanclothes.org/resources/publications/italian-living-wage-report/. Accessed July 2016
Coach (2016) Sustainability report. http://www.coach.com/sustainability-report.html. Accessed July 2016
Corporate Human Rights Benchmark (2016) Pilot benchmark. https://business-humanrights.org/en/corporate-human-rights-benchmark-0/2016-pilot-benchmark. Accessed July 2016
Ernest & Young (2015) Seeking sustainable growth – The luxury and cosmetics financial factbook. http://www.ey.com/Publication/vwLUAssets/EY_Factbook_2015/$FILE/EY-Factbook-2015.PDF. Accessed July 2016
Ethical Trading Initiatives & Hult Business School (2016) Corporate approaches to addressing modern slavery in supply chains. http://www.ethicaltrade.org/sites/default/files/resources/corporate_approaches_to_addressing_modern_slavery.pdf. Accessed July 2016
Ferragamo S (2016) Financial reports. http://group.ferragamo.com/en/investor-relations/financial-documents/2015 . Accessed July 2016
Hermès (2016) Annual report. http://finance.hermes.com/var/finances/storage/original/application/630394b9a95603bf81b399744dbbf044.pdf. Accessed July 2016
ILO (1998) ILO declaration on fundamental principles and rights adopted by the international labour conference at its eighty-sixth session, Geneva, 18 June 1998 (Annex revised 15 June 2010)
Joy A, Sherry J Jr, Venkatesh A, Wang J, Chan R (2012) Fast fashion, sustainability, and the ethical appeal of luxury brands. Fash Theory 16:273–295. https://doi.org/10.2752/175174112X13340749707123
Kantz C (2007) The power of socialization: engaging the diamond industry in the Kimberley process. Bus Polit 9:1–20. https://doi.org/10.2202/1469-3569.1186
Kering (2016) Sustainability report. http://www.kering.com/en/sustainability/stakeholders-reporting#anchor0. Accessed July 2016
KnowTheChain (2016a) KnowTheChain benchmarking 2016 via https://business-humanrights.org/en/knowthechain-benchmarking-2016. Accessed July 2016
KnowTheChain (2016b) Apparel and footwear company disclosure. https://business-humanrights.org/en/knowthechain-apparel-and-footwear-company-disclosure. Accessed July 2016
Kors M (2016) Annual reports. http://investors.michaelkors.com/financials/annual-reports/. Accessed July 2016
Lauren R (2016) Citizenship report 2015 via https://drive.google.com/file/d/0B5wkVXqXJ0GDQTZFMHlpcFN2dkk/view?usp=sharing. Accessed July 2016
Likoksy M (2008) Gender arbitrage: law, luxury and labor. Wis J L Gend Soc 23:293
LVMH (2016) Code of conduct. https://www.lvmh.fr/wp-content/uploads/2014/11/code-of-conduct.pdf. Accessed July 2016
OECD (1976/2016) OECD declaration and decisions on international investment and multinational enterprises
Prada (2016) Code of ethics. Financial reports. http://csr.pradagroup.com/content/uploads/2015/06/code_of_ethics_en.pdf , http://www.pradagroup.com/en/investors/financial-reports. Accessed July 2016
Project Just (2016) Brand profiles. http://projectjust.com/. Accessed July 2016
RobecoSAM (2016) Sustainability indices. http://www.sustainability-indices.com/. Accessed July 2016
Ruggie J (2013) Just business: multinational corporations and human rights. Norton, New York
Shift and Mazars (2016) UNGP reporting database company page of LVMH. http://www.ungpreporting.org/reportingdatabase/companies-page/. Accessed July 2016
Sun H (2013) Living together in one civilized world: how luxury companies and consumers can fulfill their ethical responsibilities to the poor. UCDL Rev 47:547
United Nations Global Compact (2016) About us. https://www.unglobalcompact.org/AboutTheGC/index.html. Accessed July 2016

United Nations Human Rights Council (2011) Guiding principles on business and human rights. Implementing the United Nations "Protect, Respect and Remedy" Framework, U.N. Doc. A/HRC/17/L.17/31. Accessed June 2011

Winston A (2016) Luxury brands can no longer ignore sustainability. Harvard Business Review, 8 February. https://hbr.org/2016/02/luxury-brands-can-no-longer-ignore-sustainability. Accessed July 2016

Melawear: Inspired by Cradle to Cradle

Henning Siedentopp

1 About Melawear

The name MELAWEAR derives from the Hindi word "mela" meaning "acting together". MELAWEAR develops and produces organic cotton textiles such as garments, bags and sneakers within Indian industrial supply chains. Inspired by the Cradle to Cradle (C2C) product standard,[1] the main material for all products is Fairtrade certified organic cotton from India.[2] The organic cotton is processed, dyed and stitched in Indian supply chains, certified in its entirety by the Global Organic Textile Standard (GOTS).[3]

India represents one third of the world's global cotton production and all production steps from raw cotton to final garments can be fulfilled in domestic plants. India represents the world's largest democracy and offers supply chains and auditing systems for all stages in cotton textile production. MELAWEAR's entire sourcing and production process from raw organic cotton to final textiles takes place entirely in India to ensure a minimum of externalities for transportation and a maximum of value creation for the country of origin. All textiles are always certified by both, the Fairtrade Cotton Standard (Fairtrade) and GOTS and sold as such within European retail shops.

MELAWEAR started to sell its first collection in Spring 2015 being listed in more than 100 European retail shops after only 12 months of operating. With the launch of the first backpacks and sneakers in the world certified by both Fairtrade and GOTS, MELAWEAR marks a key milestone in the field of sustainable textiles, not yet

[1]Cradle to Cradle Certified™ Product Standard, Version 3.1 (2014).

[2]Fairtrade Standard for Fibre Crops for Small Producer Organizations (2011).

[3]Global Organic Textile Standard, Version 4.0 (2014a, b).

H. Siedentopp (✉)
Melawear GmbH, Lüneburg, Germany
e-mail: info@melawear.de

S. M. Jastram, A.-M. Schneider (eds.), *Sustainable Fashion*, Management for Professionals, https://doi.org/10.1007/978-3-319-74367-7_10

achieved by any other fashion company. For its success, MELAWEAR was awarded several times. Most important is the 2015 announcement by the European Commission to be among the top ten European social innovators (European Commission 2015) and to become a finalist of the entrepreneurship award for sustainability (Gründerpreis Nachhaltigkeit 2016). Additionally, the state of Lower Saxony awarded the CEO of MELAWEAR as the student scientist of the year for his research and study results in the field of transforming theoretical knowledge on sustainability into practical actions.

MELAWEAR unites entrepreneurial simplicity and a holistic sustainability approach in the core of its business. Selling plain and basic garments, bags and sneakers, produced in high volumes, as a never out of stock collection that is always available, characterises MELAWEAR's business model. Therefore, MELAWEAR does disrupt the current fashion industry in several ways and succeeds in this attempt by

1. Ignoring established habits and procedures such as fast changing fashion trends, seasonal adjusted collections, or pre-order presentation of new collections and
2. Avoiding or reducing any marketing expenditures, financial expenditures are focused on all aspects of sustainable product design and production.
3. By reducing the complexity for everybody in the supply chain from the designer, via the producer up to the retailer, MELAWEAR offers the option to sell sustainable textiles at a competitive quality, design, and price in the conventional textile market.
4. From the very beginning, all savings that have come from MELAWEAR's holistic business model have been used to improve the product and environmental impact, to reach higher social standards along the supply chain, to keep the prices low as well as to invest in research and development for improvements at all stages.

The given business case for sustainability will focus on how the C2C standard guides MELAWEAR along its corporate strategy. It shows the understanding and integration of sustainability as a main part of MELAWEAR's business strategy within current global textile challenges. Applied sustainable textile standards, such as Fairtrade and GOTS, will be explained with a focus on the C2C standard, in order to show the main source of inspiration for MELAWEAR's sustainability approach.[4] Therefore, an explanation on what C2C is about, how it is applied, and the resulting challenges and opportunities for MELAWEAR, will be discussed.

[4]An unpublished Master-Thesis from the author of this article focuses on the comparison of the Fairtrade, GOTS and C2C standard and acts as a guideline for this article.

2 Sustainability and Textile Industry

2.1 Global Textile Challenge

The sustainability approach of MELAWEAR and its business strategy are interdependent to current global social, environmental, and political challenges that are numerous and part of daily news: an increasing population growth of 30% towards 9.5 billion people until 2050 (Statista 2015a), 800 million people suffering from hunger (Food and Agriculture Organization of the United Nations 2015), 1.2 billion people that have to live on less than $1.25 a day (United Nations 2015a), and 50 million people that are on the run from either hunger, war, or poverty (UNHCR 2014) are just few of the significant examples of current global challenges.

Textiles, the second largest consumer good in Germany (Statista 2015b), and its global industry are part of maintaining these social inequalities as well as environmental destruction. Cotton production has the biggest share by making 38% of all textiles produced globally (Grose 2009: 33; Piegsa 2010: 6). Even though only 2.4% of the world's agricultural land is cultivated with cotton, 11% of the worldwide consumption of toxic pesticides and 25% of harmful insecticides are used for cotton cultivation (ibid.). According to the Greenpeace DETOX campaign more than two thirds of Chinese waters are contaminated with harmful chemicals, mainly due to wastewater from the textile industry (Greenpeace 2014). The so-called "Sumangali" principle, mainly known from Indian cotton spinning mills, keeps women working as slaves in order to earn their endowment for their wedding (Exner 2013). The collapse of Rana Plaza in 2013 with the death of more than 1100 Bangladeshi is one of the most recent and prominent disasters within the industrial textile industry (Clean Cloth Campaign 2015).

Several political actions towards a more sustainable textile industry, relevant for European companies working in global supply chains, have been launched over the last decade. These include the REACH regulation (2007) by the European Union (EU) focussing on the use and restriction of chemicals applied (ECHA 2007), the German Textile Partnership (2015) as a multi-stakeholder initiative to improve sustainability along the entire textile supply chain (The Textiles Partnership 2014), the Landmark Project (LANDMARK project 2011) that has lead into an EU directive for socially responsible public procurement (2016) (European Commission 2016) as well as the seventeen Sustainable Development Goals (2016) from the United Nations (UN) focusing on social, environmental and economic equality (United Nations 2015a, b) are just some of the relevant political actions relevant for global textile challenges. The sustainability strategy applied by MELAWEAR was derived from the context of the given challenges as well as theoretical approaches, and shall be explained in detail below.

2.2 Sustainability Strategies

For MELAWEAR, the term sustainability is used as a synonym for sustainable development in order to show that sustainability is more of a process rather than a final state. The balance of social, ecological, and economic dimensions in any stage of business practice characterises MELAWEAR's corporate sustainability culture. As a small and medium-sized enterprise MELAWEAR can be described as an "emerging David" trying to show how sustainable business innovations can be launched and brought into practice (Hockerts and Wüstenhagen 2010). Three main strategies for sustainability are being discussed in the literature and applied by companies: efficiency, sufficiency, and consistency (Grunwald and Kopfmüller 2012: 92). All three strategies will be explained briefly in order to show how and why MELAWEAR applies them all by focusing on consistency as its core strategy for sustainability.

The efficiency strategy dates back to the beginning of the global sustainability debate in 1992 during the UN Rio conference. The strategy focuses on the reduction of material and energy consumption by maintaining existing quality and output (Huber 2000: 110). According to this strategy efficiency within the production process should reduce the consumption of materials and energy by ten times (Grunwald and Kopfmüller 2012: 93). Textiles with less or no cutting waste derived from adjusted designs and patterns, are a good example for efficiency strategy.

The sufficiency strategy focuses on the consumer by claiming that a reduction in consumption will lead to a reduction of environmental problems (e.g. pollution) due to a lower use of resources (Michelsen 2012: 89). Additionally, a reduced consumption would lead to a reduced need of resources (e.g. income) and therefore an increase in time available for self-supply (Paech 2011: 10–11). Textiles that are sold and purchased within the second-hand market are a good example for the sufficiency strategy as several consumers are using one textile.

The consistency strategy is inspired by nature transferring the functionality of biosphere into production and consumption processes. By providing closed cycles of materials, waste and pollution are being avoided or entirely reused within biological or technical cycles (Braungart and Mc Donough 2013). Textiles that can biodegrade after their usage are a good example for consistency strategy as they become nutrients in the soil.

The consistency strategy differs from the other two strategies as the focus is based on qualitative, instead of quantitative, resource usage. An optimization of the efficiency in any production process leads to sustainable achievements, but only to a certain extent (Huber 2000: 5–6). For the given example, while reducing the cutting waste it remains unclear if the textile contains toxic chemicals or if the production is run under good social standards. By lowering the consumption of resources within a sufficiency strategy the time will be extended until natural resource reserves are vacant. For the example of a second-hand textile it would mean that we can extend the life of textiles; they will still have an end of life where resources will be wasted. Both strategies therefore show their limits by having a focus based on quantitative aspects. MELAWEAR applies all three strategies but focuses on the consistency strategy with a clear focus on qualitative aspects in its sustainability strategy.

3 Melawear's Sustainability Strategy

The foundation of MELAWEAR is directly linked to the environment of the Leuphana University in Lüneburg, Germany, an environment characterized by its strong focus on sustainability and entrepreneurship, both in academics as well as practical research. The inclusion of the latest scientific expertise combined with a knowledgeable business network plays an important role in the sustainability strategy of MELAWEAR. This includes an on-going analysis of existing business solutions for sustainable textiles ready to be integrated. Furthermore it combines the ability to find out how these existing business solutions can be put together in order to make them work within the business model of MELAWEAR.

One main strategic success factor for MELAWEAR is the reduction of complexity within the business model, in order to follow a clear focus on sustainability. The decision on the right product portfolio aligning with a consistency strategy is the starting point. A special focus on the materials and accessories used for any MELAWEAR textile represents the main work of day-to-day business when it comes to an application of the chosen strategy. From the very beginning all partners within the supply chain have been selected carefully regarding their ability and willingness to extend their business practices towards consistency. Personal, professional, and long-term oriented relationships built the ground for an essential knowledge exchange within transparent supply chains and processes. This includes relationships beyond the first and second tier up to sub-suppliers for raw, auxiliary, and operating materials.

In contrast to other sustainable textile companies MELAWEAR has a clear commitment to not only stick to its sustainability standards while growing, but also to expand them equally on all three pillars for social, environmental and economic aspects. In order to ensure this, MELAWEAR adheres to the highest, international product standards for the entire product portfolio. Therefore, all textiles are certified by Fairtrade and GOTS. According to scientists and NGOs, the combination of both standards can guarantee highest sustainability practices in natural fibre production (Hansen and Schaltegger 2013; GET CHANGED 2013). By integrating globally applied and established sustainability standards into the business model, MELAWEAR can scale its business model into the mass market. The orientation along the C2C Product Standard for all MELAWEAR textiles adds even more value by going beyond Fairtrade and GOTS in environmental quality as well as life cycle perspective. Fairtrade and GOTS will be explained briefly before the C2C standard and its application at MELAWEAR will be described.

3.1 Fairtrade Cotton Standard

MELAWEAR cooperates with Fairtrade International by using organic Fairtrade Certified Cotton in order to actively improve the working and living conditions for the cotton farmers in India. This standard ensures the commitment to highest social standards and includes the prohibition of any child or forced labour (ILO criteria), the right for trade unions, the payment of minimum wages as well as social benefits, such as regulated day leaves, clean drinking water, and regulated idle times. The payment of a fixed price for cotton plus an additional bonus for the farmer's cooperatives particularly connects economic value with social benefits. Fairtrade cotton farmers are independent from global cotton price fluctuations as the price for Fairtrade cotton is fixed and adjusted only every 2 years. Fairtrade makes sure that the higher price is paid- either Fairtrade or global cotton price. An additional bonus is guaranteed to any Fairtrade farmer for its cotton with an additional premium on organic cotton. Moreover, Fairtrade actively fights poverty through direct investment in education and health (Fairtrade Trader Standard 2015; TransFair 2008, 2014).

The economic system of Fairtrade Cotton for the farmers has been extended to the entire supply chain with the new Fairtrade Textile Standard launched in the beginning of 2016. The new standard seeks to improve working and living conditions of all workers within Indian textile supply chains. The tailors are especially in focus as their wages will rise from minimum to living wages over 6 years under the new standard. This increase will more than double the current wages of tailors in India. MELAWEAR is one of the first companies participating in the new Fairtrade Textile Standard as a pilot partner. The described cooperation with Fairtrade International allows mela wear to build up lasting and trustful trade relations with farmers and producers. In turn, this guarantees long term financial security for any worker in the supply chain.

3.2 GOTS

The Global Organic Textile Standard (GOTS) for natural fibres, such as organic cotton, focuses on farming and environmentally sound textile production for the entire supply chain. GOTS helps MELAWEAR establish the highest protection for employees and the environment throughout the complete supply chain. The working conditions of the farmers are improved due to organic cotton farming where toxic fertiliser or pesticides are forbidden. The soil is protected and no human being—neither the workers nor the consumers—has contact with toxic chemicals. Every participant in a GOTS approved supply chain must have an environmental management system in place. All aspects from farming, via dying and printing to warehousing and packaging, are covered and regulated under GOTS. Regular audits are conducted for GOTS approved participants in the supply chain to ensure that the standard is being applied correctly (GOTS 2014a, b). This is how MELAWEAR succeeded at delivering healthy, environment-friendly and sustainable textiles to the consumer from day one.

Farmers are only allowed to cultivate organic cotton. Specific lists ban or limit chemicals to certain values and do apply to any fabric and garment certified by GOTS. Examples of banned chemicals for GOTS textiles are aromatic and halogenated solvents, formaldehyde and other short-chain aldehydes, heavy metals as well as genetically modified organisms. Through banning such chemicals the environment is protected during the dying and processing stages of organic cotton fabrics (GOTS 2014b, 2015). GOTS certified textiles from MELAWEAR are tested in German and Indian laboratories on a regular basis to ensure that they do not contain any restricted or banned toxic chemicals. The blend of organic cotton fabrics with non-natural fibres is allowed under GOTS for up to 30% of the garment and must be marked clearly on the garment. MELAWEAR does not blend its fabrics but sticks to the usage of 100% organic cotton for two reasons: First, to prepare garments the become biodegradable and secondly for making new cotton out of used garments (more details in the next chapter about C2C).

3.3 *Cradle to Cradle*

While the Fairtrade Cotton Standard and GOTS focus on natural fibre textiles only, the concept of C2C comprises a more holistic product standard, not only specialised in textiles or natural fibres. In contrast to Fairtrade and GOTS, the C2C Product Standard can be applied to any product by setting a practical framework for any consistent sustainable product design. Inspired by nature C2C does not acknowledge any form of waste or abstinence.

The C2C design approach is built around three main principles (Cradle to Cradle Certified™ Product Standard 2014):

1. *Waste equals food* describes the principle observed in nature where no waste exists. Any kind of waste in a man-made production, may it be during production process or after the usage of a product, becomes either a biological or technical nutrient for new production. In this way waste will be either technically reused for upcoming productions or organically composted to become fertiliser for new cultivations. As an example for the sustainable textile industry this could mean that used and old garments can either be converted into new garments or composted in order to grow new cotton.
2. *Use of renewable energy* describes the principle that renewable energy should be used within the production and transportation process instead of fossil fuels and non-renewable resources. Solar, wind, water, and biomass energy are ways to harness energy in order to run production facilities sustainably instead of using oil, gas, or coal as sources of energy. As an example, for the sustainable textile industry this could mean that the energy for production facilities for tailoring

would be generated through solar panels, and biogas installations could be used for the heating of dying units.

3. *Celebrate diversity* describes the principle in which products and services created must not follow any standardized form or established procedure. Rather new and innovative ways of combining and designing products should be in place as a creative design approach. As an example, for the sustainable textile industry this means to bring in new materials or new material compositions as well as to create processes that have not existed or been used in that way previously.

Traditionally, industrial product design can be described as "Cradle to Grave", where products do have an end of life cycle without further consideration of a further usage of their components. In contrast to this traditional design approach, the C2C design approach distinguishes between two continuous cycles, the technical and the biological cycle (Cradle to Cradle Certified™ Product Standard 2014).

The technical cycle for materials does apply for any kind of primary materials or resources such as metal or plastic. As the demand for such scarce resources is rising they should circulate within cycles. When it comes to textiles, such as garments or accessories, metal zippers, buckles, or buttons are good examples of components for a technical cycle. The *biological cycle* for materials applies to any materials that are environmentally-friendly and compostable. Through composting after usage, such materials become nutrients for the soil and therefore the starting point for new resources. Before materials become composted, multiple usages can be conducted in cascades, where the quality for reusing materials decreases in each step of usage. When it comes to textiles, the fabric of a t-shirt, made out of organic cotton, coloured and stitched with compostable dyestuff and sewing yarn, is a good example for the biological cycle (Cradle to Cradle – Wiege zur Wiege e.V. 2015).

Any product that is designed and produced inspired by the C2C standard can become certified. As the standard does have five classification levels, it offers the opportunity for any company to evaluate its compliance within the C2C standard. Starting with BASIC, the *C2C certificate* goes up to BRONZE, SILVER, GOLD and PLATINUM. The Products Innovation Institute in the United States remits the corresponding level for each product after an individual certification process. The C2C standard is very comprehensive, detailed and specific by offering a holistic approach for sustainable product design. To evaluate a product according to the C2C product standard, designers and manufactures are guided through a continuous improvement process that evaluates a product through five categories briefly explained as follows (Cradle to Cradle Certified™ Product Standard 2014; Cradle to Cradle Products Innovation Institute 2016):

1. *Material Health* is about knowing the chemical ingredients of every material in a product, and optimization towards healthier materials. It involves the process of identifying materials as either biological or technical nutrients. It is furthermore essential to understand how chemical hazards combine with likely exposures to determine potential threats to human health and the environment during production, usage and after usage.

2. *Material Reutilization* is about designing products so that the materials used can either be returned to nature or into industry processes. It includes the ability of a designer to maximize the percentage of renewable materials or recycled content used in a product. Additionally, it is important to increase the percentage of materials that can be safely reused, recycled, or composted when the product is no longer used.
3. *Renewable Energy & Carbon Management* is about increasing the amount of renewable energy used within the production process up to 100%. Renewable energy such as solar power, water, and wind energy should be used as source of energy and carbon emissions arising during production should be counteracted.
4. *Water Stewardship* is about recognizing the right to clean and sufficient water for anybody worldwide. Further it should address local water impacts at each manufacturing facility and industry. By keeping track of the amount of water used within the production and by avoiding water waste or pollution, the production process should be optimized. Any harmful chemicals used within the production process must be identified, separated, and filtered so the effluent stays clean for environmental usage.
5. *Social Fairness* is about the valuation of all people and natural systems affected by the creation, use, disposal or reuse of a product. Using globally recognized resources and third party audits to conduct self-assessments for identification of local and supply chain issues are core parts of C2C to assure optimal social conditions.

4 Application, Challenges and Opportunities

The concept of C2C is the driving force for MELAWEAR when selecting the right business partner in a knowledge driven company culture. C2C formed the basis from day one, consisting of an on-going screening and integration from both scientific as well as practical research results. MELAWEAR is actively searching, analysing, evaluating and visiting its current and future business partners in India. Extensive field trips have been used to discuss and explain the concept of C2C and its application within the textile industry. The willingness and readiness of textile producers in India were key features for selecting the right partners. Their knowledge and understanding of the C2C concept was as important as their current status quo: the usage and quality of organic materials, the percentage of renewable energy used and the presence of a carbon management system, the quality of water stewardship and effluent plants as well as social standards and practices applied were dominant factors when selecting MELAWEAR's business partners. The compliance with Fairtrade and GOTS is the minimum requirement for any MELAWEAR partner. Conventional aspects of supplier selection such as low purchase prices or short delivery lead-time have been considered but have not been leading factors in the selection of business partners.

The design and production process at MELAWEAR focuses on the ability of any textile or accessory to be kept within a biological cycle with a technical slope. As organic Fairtrade Cotton is the main material for MELAWEAR textiles, the compliance with the biological cycle of C2C is the first goal of MELAWEAR. As accessories are in place in many garments, MELAWEAR designs textiles in a way that accessories are either biodegradable or technically reusable. Such technical slopes enable the option of material reutilization of the organic cotton used in MELAWEAR textiles in order to make new yarn, fabric, and finally textiles from it. For that reason, MELAWEAR textiles are never mixed nor blended but are always homogeneously made out of organic Fairtrade Cotton only.

Chemical and material analyses at independent laboratories are essential in order to ensure material health for MELAWEAR garments according to GOTS and C2C. MELAWEAR therefore selects materials based on test results and optimizes them for the organic cycle by avoiding materials that are not biodegradable. Whenever available C2C certified materials are being integrated into the supply chain and production process. New materials which are not yet certified by GOTS or C2C are tested and compared with GOTS and C2C standards before used. Any materials used must undergo chemical analysis before it can be used in a MELAWEAR textile sold to the end customer.

The communication of MELAWEAR's sustainability strategy, which is inspired by C2C, was shared with its customers from the very beginning. By explaining how social, environmental, and economic aspects have been integrated into the business model, a transparent customer communication was established. As MELAWEAR started as a new market participant without any history in industry, no justifications about previous business practices where needed. The compliance with Fairtrade and GOTS have been made visible and linked into MELAWEAR's communication. The orientation of the C2C standard is communicated to customers in order to show MELAWEAR's vision.

Due to the innovative approach of applying C2C within Indian textile supply chains MELAWEAR has to face some challenges. Transferring existing knowledge of sourcing or producing sustainably is as complicated as finding right ways of measuring or analysing materials or energy usage. Additionally, only a few materials and suppliers are currently certified by C2C or are available with only limited access. Internal resources at MELAWEAR are limited too, as it is a new SME with limited access to time, knowledge, and capital. Whereas some textiles are predominated for the C2C design approach, others are not due to different materials or functions needed.

Nevertheless, MELAWEAR's sustainable business model, inspired by C2C, is an opportunity to achieve a competitive advantage. Achieving the application of sustainability knowledge from Fairtrade organic cotton farmers, material producers, dying units, production plants up to customers, MELAWEAR becomes a flagship project for a sustainable textile industry. By enabling the textile industry to support a world without waste, a healthy environment and social equality are created. MELAWEAR, together with its partners and customers will create a positive footprint. Making sustainable fashion available for everyone is therefore no longer a goal but part of reality driven by MELAWEAR.

References

Braungart M, Mc Donough W (2013) Intelligente Verschwendung – the upcycle: Auf dem Weg in eine neue Überflussgesellschaft. Oekom-Verlag, München

Clean Cloth Campaign (2015) Rana plaza: a man-made disaster that shook the world. http://www.cleanclothes.org/ua/2013/rana-plaza. Accessed 15 June 2016

Cradle to Cradle – Wiege zur Wiege e.V (2015) Der Biologische Kreislauf, Der technische Kreislauf. http://c2c-ev.de/c2ckonzept/kreislaeufe/. Accessed 18 June 2016

Cradle to Cradle Certified™ Product Standard, Version 3.1 (2014) http://s3.amazonaws.com/c2c-website/resources/certification/standard/C2CCertified_ProductStandard_V3.1_160107_final.pdf. Accessed 15 June 2016

Cradle to Cradle Products Innovation Institute (2016) Product certification. http://www.c2ccertified.org/get-certified/product-certification. Accessed 18 June 2016

ECHA (2007) Understanding REACH. http://echa.europa.eu/regulations/reach/understanding-reach. Accessed 16 June 2016

European Commission (2015) Announcing the 2015 European social innovation competition finalists. http://ec.europa.eu/growth/tools-databases/newsroom/cf/itemdetail.cfm?item_id=8520&lang=en&title=Announcing-the-2015-European-Social-Innovation-Competition-finalists. Accessed 17 June 2016

European Commission (2016) Supporting social responsibility in the economy through public procurement. http://ec.europa.eu/growth/tools-databases/newsroom/cf/itemdetail.cfm?item_id=8667&lang=en&title=Supporting-social-responsibility-in-the-economy-through-public-procurement. Accessed 15 June 2016

Exner M (2013) Sumagali-System- "Knochenarbeit – zehn, oft zwölf Stunden am Tag", Published in DIE ZEIT. http://www.zeit.de/wirtschaft/2013-12/textilindustrie-kinderarbeit-sumangali-system-in-indien. Accessed 15 June 2016

Fairtrade Standard for Fibre Crops for Small Producer Organizations (2011) http://www.fairtrade.net/fileadmin/user_upload/content/2009/standards/documents/2011-10-15_EN_SPO_Fibrecrops_Standard.pdf. Accessed 15 June 2016

Fairtrade Trader Standard (2015) Current version: 01.03.2015 v1.0. http://www.fairtrade.net/fileadmin/user_upload/content/2009/standards/documents/generic-standards/2015-03-01_TS_EN.pdf. Accessed 15 June 2016

Food and Agriculture Organization of the United Nations (2015) The state of food insecurity in the world. http://www.fao.org/3/a-i4646e.pdf. Accessed 15 June 2016

GET CHANGED! (2013) Bewertung Nachhaltigkeitsstandards im Bekleidungs- und Textilsektor. http://www.getchanged.net/de/magazin/hintergrund/glaubwuerdig-und-effektivkonsumentenschutz-organisation-empfiehlt-vier-labels-fuer-faire-mode-4073.html?page. Accessed 17 June 2016

GOTS (2014a) http://www.global-standard.org/the-standard/general-description.html. Accessed 18 June 2016

GOTS (2014b) Global Organic Textile Standard (GOTS) – Version 4.0. http://www.global-standard.org/images/GOTS_Version4-01March2014.pdf. Accessed 18 June 2016

GOTS (2015) Compare – from field to fashion! http://www.global-standard.org/images/stories/GOTS_Flyer_Compare121114.pdf. Accessed 18 June 2016

Greenpeace (2014) Textilindustrie vergiftet Gewässer. https://www.greenpeace.de/themen/endlager-umwelt/textilindustrie-vergiftet-gewasser. Accessed 15 June 2016

Grose L (2009) Sustainable cotton production. In: Sustainable textiles, life cycle and environmental impact. Woodhead, Cambridge, pp 33–62
Gründerpreis Nachhaltigkeit (2016) http://www.gruenderpreis-nachhaltigkeit.de. Accessed 17 June 2016
Grunwald A, Kopfmüller J (2012) Nachhaltigkeit, 2nd edn. Campus Verlag, Frankfurt am Main
Hansen EG, Schaltegger S (2013) 100 per cent organic? A sustainable entrepreneurship perspective on the diffusion of organic clothing. Corp Gov Intern J Bus Soc 13(5):583–598
Hockerts K, Wüstenhagen R (2010) Greening Goliaths versus emerging Davids – theorizing about the role of incumbents and new entrants in sustainable entrepreneurship. J Bus Ventur 25: S.481–S.492
Huber J (2000) Industrielle Ökologie. Konsistenz, Effizienz und Suffizienz in zyklusanalytischer Betrachtung. Global change. Arno Spitz Verlag, Berlin, pp 109–126
LANDMARK project (2011) The LANDMARK project – moving towards socially responsible public procurement. http://www.landmark-project.eu/en/about-landmark. Accessed 15 June 2016
Michelsen G (2012) Grundlagen einer nachhaltigen Entwicklung. Centre for Sustainability Management, Leuphana Universität, Lüneburg
Paech N (2011) Vom grünen Wachstumsmythos zur Postwachstums Ökonomie. In: Welzer H/Wiegandt K (Hrsg.) Perspektiven einer nachhaltigen Entwicklung. Fischer Taschenbuch Verlag, Frankfurt, pp 131–151
Piegsa E (2010) Green fashion – Ökologische Nachhaltigkeit in der Bekleidungsindustrie. Diplomica Verlag, Hamburg
Statista (2015a) Prognose zur Entwicklung der Weltbevölkerung von 2010 bis 2100 (in Milliarden). http://de.statista.com/statistik/daten/studie/1717/umfrage/prognose-zurentwicklung-der-weltbevoelkerung. Accessed 15 June 2016
Statista (2015b) Private Ausgaben fürKonsumgüter in Deutschland nach Segmenten in den Jahren 2009 bis 2013 (in Milliarden Euro). http://de.statista.com/statistik/daten/studie/199403/umfrage/privateausgaben-fuer-konsumgueter-nach-konsumgueterbereichen/. Accessed 15 June 2016
The Textiles Partnership (2014) About us. https://www.textilbuendnis.com/en/home/about-the-textile-partnership. Accessed 18 June 2016
TransFair (2008) Fairtrade-Baumwolle – Ein Gewinn für alle. http://docplayer.org/1768691-Fairtrade-baumwolle-ein-gewinn-fuer-alle-www-transfair-org.html. Accessed 18 June 2016
TransFair (2014) Baumwolle – Produktinformation von TransFair 2014. https://www.fairtrade-deutschland.de/fileadmin/DE/mediathek/pdf/fairtrade_baumwolle_factsheet.pdf. Accessed 18 June 2016
UNHCR (2014) Über 50 Millionen weltweit auf der Flucht. http://www.unhcr.de/home/artikel/2cadec7ad82a795cc394fde14cfe54f0/ueber-50-millionen-weltweit-auf-der-flucht.html. Accessed 15 June 2016
United Nations (2015a) We can end poverty. http://www.un.org/millenniumgoals/pdf/Goal_1_fs.pdf. Accessed 15 June 2016
United Nations (2015b) Sustainable development goals kick off with start of new year. http://www.un.org/sustainabledevelopment/blog/2015/12/sustainable-development-goals-kick-off-with-start-of-new-year/. Accessed 15 June 2016

From Hospital Sheet to High Quality Design Dress: Tauko's Upcycling Method to Sustainable Fashion

Mila Moisio

TAUKOdesign is a creative design company that owns the sustainable fashion label TAUKO. The company is based in Helsinki and is owned and run by the founders and designers Mila Moisio and Kaisa Rissanen (TAUKO 2017).

We focus on the design process, refinement of used textiles, and locally produced high quality womenswear collections. We provide solutions for production and textile development of ecological fashion. Our production is based on a network of regional partners with suppliers in Finland and Germany.

Discarded service textiles are the raw materials for high quality womenswear collections.

TAUKO womenswear collections are made from otherwise discarded textiles that are carefully sorted and selected. To guarantee continuous material flow for the production, we work in close collaboration with Finnish and German textile service companies. Their core field is renting textiles used in health care, hotels, and restaurants. Around 30–60% of all of their textiles are flat, such as bedlinen or table clothes. They are made to withstand heavy use, yet feel silky and soft. Over their extensive use within this first life, their quality diminishes in certain areas: permanent stains appear, and the textile wears out at certain parts.

Such textiles cannot be used by hotels, restaurants, and health care systems anymore and are therefore considered waste by the textile maintenance companies. This applies as well to out-fashioned materials that feature e.g. old design elements. Tons of these textiles are discarded every year, resulting in waste of high quality materials. For parts of this waste, TAUKO provides an alternative to the landfill by giving it a new life as the perfect raw material for TAUKO collections.

M. Moisio (✉)
TAUKO Design, Helsinki, Finland
e-mail: tauko@taukodesign.fi

S. M. Jastram, A.-M. Schneider (eds.), *Sustainable Fashion*, Management for Professionals, https://doi.org/10.1007/978-3-319-74367-7_11

We sort the used textiles and select the ones that have sufficient quality for further high-quality refinement. TAUKO thereby offers a production and service model that radically lowers the amount of textile waste.

Due to the production that is built on the recycling of old textiles, no virgin materials are used for TAUKO collections. The raw materials are altered according to our designs, while aiming to use existing infrastructure and machinery of the textile industry. Most of the material is re-dyed in seasonal colours, silk printing is used to add colourful prints to the collections, and all textiles of outdoor designs are coated to provide water resistance.

1 Starting to Explore the World of Used Textiles

The idea to use old textiles for fashion has been the driving force of the company since its foundation in 2008. We both studied art research and, after finishing our MAs, we founded the creative design company that later became TAUKOdesign.

Initially, we searched through vintage shops, flea markets, and storage units of theatres for old curtains, bed linen, and table clothes to be used for making our designs. While experimenting with different vintage textiles we soon discovered the quality differences between textiles produced before and after the 2000s. The older the textile, the better the quality, usually. It was very interesting to discover the high quality of old, used vintage textiles and how well they fitted for fashion. But time and time again we were disappointed by the quality of the used textiles that we produced during the last 15 years. The difference in the quality of the fabrics was remarkable. The textiles we were looking at were mostly household textiles like curtains and bedlinen. After our survey through second-hand shops and flee markets, the search for high quality recycled textiles became one of our main missions.

Starting with vintage home textiles, the brand first focused on producing unique pieces, highly customized for individual end user clients. As the business grew, it demanded for bigger scale production and at least small batches of our designs. We aimed for a collection-based business model following the production and sales structure of the fashion market. Therefore, we needed to find a source for high quality textiles with regular supply.

We started working with Finnish textile maintenance services in 2009 to look into textiles they discard from their service. The first companies that we worked with, provided textile maintenance for the health care sectors in Helsinki area. The products that the companies provided for their customers were made to withstand frequent washing, drying, and mechanical treatment. They had kept the same high quality standards during the past decades in contrast to what we experienced in the home textiles made for individual consumers. By developing a series of sample collections, we found those textile products that were best suited for our production and that satisfied our demand for high quality. The first womenswear collection made completely from these recycled health care textiles was released in 2013. After our research for the first phase of the development of our production method, we have been able to develop and apply the same method to our latest collections. Finding the right materials and ways to work with them have been key factors in developing our concept.

In 2015, an MA student from Tampere Technical University wrote his master's thesis on material sciences about our used materials. The results indicated that the textiles used by TAUKOdesign were of high quality and suitable for fashion production (Vats 2015). This proved what we assumed: the discarded service textiles were more than suitable as a high-quality and continuously available material for our fashion line of sustainable womenswear.

At the same time as we researched the materials, we extended our material supply by collaborating with German textile maintenance and leasing companies whose major clients are hotels and restaurants.

2 Finding Solutions for Textile Waste Recycling and Upcycling

Numerous studies show how the increase of textile waste has become an unresolved problem in European countries. Shameek Vats points out how the usage of textiles per capita rose by 40% during the past 10 years. This led to a lot of questions in the public about the amount of consumer textile waste. The EU regulations have taken this into account by forbidding any organic waste, including organic textiles, in landfills in order to prevent the increase of waste that is already currently difficult to handle (Vats 2015). The production model of TAUKO offers a partial and local solution to the problem of service textile waste. Around 30% of the textile waste of the textile maintenance companies can be re-used for the production of TAUKO collections. The textiles are selected by textile service companies and TAUKOdesign.

The discussion around textile waste has introduced many different approaches to the waste problem that we are facing and therefore it has also broadened the concept of recycling. Recycling can be done in different ways and it is the most commonly used term to describe any kind of re-use of old products or materials. The terms *upcycling* and *downcycling* were established to differentiate qualities of recycling.

TAUKO's production method and concept of re-using the discarded textiles from textile maintenance companies can best be described by the term upcycling. It describes the *process* of re-using materials or products that increases the value of the used materials. In contrast to recycling, upcycling always implies the idea of added value.

The discarded textiles of the textile maintenance companies are considered as waste. To reduce the amount of textile waste, the textile service companies are developing methods for recycling with TAUKOdesign. If not used for TAUKO collections, most of them are used for energy waste or they are taken to the landfills, which cannot be considered as recycling at all.

Luckily, there are also other options than landfills or burning, as the process of mechanical recycling has been developed over the last several years. This means that the textile products are mechanically processed so that they can be used for making new products such as different liquid absorbing mats and tissues. This is an important business model that is developed around the usage of textile waste. All kind of textile products, in addition to consumer waste, can be used for these products, which makes the model efficient and sustainable. This form of recycling can be described as downcycling as the products do not maintain or add value of the recycled items.

The mechanical recycling of the textile products is efficient when it comes to production of liquid absorbing mats and tissues. For these products all kind of textiles can be used: mixed fibers as well as 100% cotton products or 100% polyester products. However, the model cannot be applied to producing textiles for fashion production as the products are very rough and do not fulfil the standards of high quality fashion garments.

Most of the discarded textiles used by TAUKOdesign consist of 50–60% cotton, 40–50% polyester. The upcycling process does not include any mechanical recycling of the textile products: the discarded products are treated as little as possible before being used for the production of TAUKO collections. The biggest work load is in the sorting process. The sorting process is carried out by TAUKOdesign and the textiles service companies. As the textiles consist of mixed fibers, it makes them difficult to recycle for fashion using the modern method of mechanical or chemical recycling. When it comes to mixed fibers in a fabric, these methods require the separation of the different fibers before they can be applied. As there is already such large amounts of easily recyclable materials, the textiles used by TAUKO are not used for mechanical and chemical recycling.

3 Working with Regional Partners

All the production phases of TAUKO collections are handled by TAUKOdesign in close collaboration with the local suppliers. The design process goes hand in hand with the textile refinement development and many of the designs are inspired by the raw materials and the specific characteristics of these textiles. The refinement of the textiles is done in collaboration with dyeing companies, university research groups and the textiles maintenance services. With every step of the operation, we keep a close eye on the ecological aspects of the production and always choose the ones that are most sustainable, ensuring the high quality and richness of the designs. Avoiding long geographical distances in all the production and distribution phases makes the production model as carbon neutral as possible. Made from 100% recycled materials, TAUKO collections are always also made locally. The raw materials are originating from Germany and Finland and the sorting is completed at the premises of the textile service companies. The sewing and cutting is completed in Finland, Estonia, and Poland: as close as possible to the source of the raw materials.

The textile maintenance and leasing companies throughout Europe offer products that are alike. By developing the method of using these discarded textiles for fashion production, TAUKOdesign is offering locally operated answers for the textile waste problem that companies are facing in different regions. The method is based on existing raw materials and industry so that production can be set up basically anywhere around Europe without building special facilities for it. The refinement of the textiles, sewing, and cutting is completed as close to the source of the raw material suppliers as possible. Using the local infrastructure of existing technology and textile industry allows us also to explore different possibilities for the production and recycling of the textiles by creating sustainable fashion. The dyeing is completed by Finnsih, Estonian and Polish companies who are commited to developing the most ecological methods available.

The concept of TAUKO collections highlights the factor of using locally discarded service textiles. As the textiles that the companies offer to their customers in Finland and in Germany are more or less the same, the same collections can be produced from local materials within regional networks. The first TAUKO womenswear collection made from German discarded maintenance textiles was launched in July 2016.

4 Facing the Challenges of a Polarized Fashion Industry

For us the biggest challenges as a young and upcoming fashion label is to understand the market that is facing a lot of changes itself. The contemporary world of fashion is more and more polarized to the extremes. One extreme is high quality sustainable clothing, while the market is dominated by fast and cheap fashion. At the same time, there is an ongoing discussion about the change, some call it collapse, of the entire retail system of fashion sales. Online platforms created a separate sale channel competing with the physical stores offering suppliers and designers direct access to the consumers without the extra costs caused by the typical retail structure.

The polarization of the fashion world and the changes that are taking place in the sale structures make it important for us to understand for which market we are aiming. In addition, we have to think of alternative ways of how our brand and concept are communicated and realized in both worlds, the established but shrinking classic sales system and the upcoming digital infrastructure. Our aim to use predominantly recycled textiles for our production further challenges us by making it mandatory to develop an individual structure for sales, production, and marketing. Social media and online sales work together as efficient channels to promote sales together with carefully selected stockists.

For us it is important to offer service, quality, and substance that is authentic and that is something we truly believe in. Our sales and marketing are currently targeted only at a small audience and do not even try to compete with the "fast fashion" world. Though there is less and less space in the busiest shopping streets for boutiques that offer high quality fashion, there is the beginning of a new trend that promotes lifestyle shops and concept stores next to big shopping malls. Instead of being the main sales channels, these small boutiques function as local showrooms that offer first impressions of the TAUKO philosophy to our future customers.

Though the market we aim for and the way we structure our sales is different from the ones of big companies providing fast fashion, their influence greatly affects how people perceive and think of fashion and clothing in general. Cheap production and fast seasonal changes create an image of fashion that it is produced without any costs. In one generation, the value of clothes has gone down with full speed and for now, there seems to be no end. The production is made as cheap as possible and the quality of the products does not really matter anymore. For a normal consumer it is hard and way too complex to understand why a dress that is made from local, recycled bedlinen is more expensive than a dress that is made from Indian organic cotton. And for us as well, it is a big and sometimes overwhelming challenge to live in a world where this is true.

5 Fashion as Sustainable Lifestyle

For us being sustainable means recognizing the environmental impact that the company's actions have on the environment. Not only the ways of production, but also the structures and policies of the sales, marketing, and design should be taken into consideration when talking about sustainability. In contemporary sustainable fashion there are a lot of different approaches, and I believe that some of these approaches can offer true answers on how to use natural resources sustainably and in balance with the environment - even creating alternative business models. The public seems to be ready for a change.

Fashion has certainly become much more than a practical necessity for bigger audiences than ever before, and the broadening of the fashion market creates new ways for sales and recreates the entire fashion market structure. International trade shows offer a selected venue for buyers, but also focus more and more on seminars and related events, as fashion is not just about the designs, trends, and the feel of silky textiles. Sustainable brands like TAUKO have a lot to offer to the demand of lifestyle brands, and hopefully this demand will bring back some small boutiques to the busiest city streets. More and more people are aware of the problems of fast fashion production and are curious to hear about new and alternative ways of manufacturing. Most of the sustainable fashion brands have a unique story to tell to the people, and these stories support the ideas of individuality and responsibility, to which the customers want to relate.

This is where a dress stops being "just a dress": beautiful looks or silky feeling on a skin, and starts to become a complex and layered story of where and how it was made and what was the inspiration behind the design. By wearing and taking care of the clothes, they become more personal and valuable and they also last longer. This is something that we see as the future of sustainable fashion: changing the way we think about fashion. Stop buying and start caring.

The polarized world of fashion offers very different approaches to fashion, and also the field of sustainable fashion is still very fragmented and mostly in the hands of small to middle size companies. We find it hard to believe that international clothing brands that have achieved certain price levels for their production would change their methods in order to guarantee fair working conditions, reasonably paid work, or non-toxic materials that would not have harmful effects on the environment. And we find it even harder to believe that the change of attitudes towards fashion would start from the big companies: they will not encourage people to mend and care for their clothes, but they will keep pushing more sales as this is how the business model prospers. The business is based on annual growth that can only be achieved by increasing sales. This means bigger production and using more resources, endlessly. Though small efforts have been made in order to make the production process seemingly more sustainable, the idea of sustainability has been applied for marketing, but real actions for truly sustainable fashion are still far out of reach.

The search for truly sustainable fashion ends up with a global economic system that is not sustainable. There is a true need for a change in our economic system that is now based on consumption and growth. As Jason Hickel (2016) pointed out "clean energy won't save us—only a new economic system will", the growth of GPD is the only measure of progress and in order to make aggregate profits the global economy needs more than 3% annual growth every year. Global material extraction and consumption has grown by 94% since 1980, and is still going up (ibid.). The big fashion companies that provide fast fashion are easily accused, but when it comes to the structure of what is actually sustainable in the long run, very few of the companies who are using ecological materials have found truly reliable sustainable bussines models.

Organic cotton is already everywhere, fair trade production has luckily also found its way to fashion, and new innovations pop up all the time in the search of sustainable fabrics. But as it is in other sectors of industry, the same applies to fashion: changing the ways of production, moving to clean energy, or giving up on fossil fuels is not enough. Cultivating cotton means cutting down trees, no matter if the cotton is grown with or without toxic fertilizers. Making a new dress means using energy, no matter how good the working conditions are. The fact is that we do not need more clothes, but we need to change the way we think about fashion and clothes. Old and recycled ones do not have less value than virgin materials and circles of the entire production need to become thought through before starting a production.

In the end we think that the key for truly sustainable fashion is in the culture of consumption: how we think about fashion and clothing. The culture of fast fashion and consumption needs to change to appreciating of the materials, valuing the things we have and making them last for longer. We need to learn again to mend and take care of our clothes. Upcycling sees the potential in a hospital sheet that can become a high quality design dress that will last for a long time and have its own story to tell.

References

Hickel J (2016) Clean energy won't save us – only a new economic system can. In: The Guardian. https://www.theguardian.com/global-development-professionals-network/2016/jul/15/clean-energy-wont-save-us-economic-system-can. Accessed 20 Jun 2017

TAUKO (2017) www.taukodesign.fi. Accessed Aug 2017

Vats S (2015) Upcycling of hospital textiles into fashionable garments. https://s3-eu-west-1.amazonaws.com/frantic/stjm/op_upcycling_hospital_textiles.pdf. Accessed 15 Jun 2017

Traceability: The Respect-Code Solution

Robin Cornelius

In 1981 I was still a student of economics and political science at the University of Lausanne, Switzerland when I founded my textile company. Having worked for the following 20 years to develop a business model compatible with ethical production, I then decided to concentrate my efforts exclusively on questions of traceability in the industry, starting with my own production. Why did I choose this direction at a time when apparently so few companies felt concerned? My conviction was that consumers should be informed in a simple and factual manner of the production steps and conditions of any purchased item. To fulfil this mission of traceability and transparency, I founded Product DNA LTD in 2005 and immediately afterwards launched www.respect-code.org. This is a public online platform which, with just one mouse click, gives access to all steps of a production process and to a description of the conditions under which garments were produced. This information is available not only to end consumers, but also to all stakeholders. I felt strongly that such transparency was sure to be welcomed by all concerned and, furthermore, that it would differentiate us from our competitors.

But is this affordable and how long will it take to achieve this vision of developing our activities in a traceable and transparent way? These were the concerns of my team. My answer was that the investment needed to implement traceability was insignificant in comparison with unforeseen costs caused by uncontrolled processes. Not to mention the negative impact of potential reputation damage due to a lack of visibility in the supply chains' daily operations. How long would it take to implement? A few days to a few weeks, depending on the number of supply chains to trace and the scope of information to be put online. From my point of view, this approach would in no way call into question our activities but could only strengthen our position on the market.

R. Cornelius (✉)
Product DNA Ltd, Lausanne, Switzerland
e-mail: robin@productdna.com

S. M. Jastram, A.-M. Schneider (eds.), *Sustainable Fashion*, Management for Professionals, https://doi.org/10.1007/978-3-319-74367-7_12

1 The Respect-Code Solution for Textiles

To date, more than 60 million products have been traced by Product DNA Ltd. The company offers traceability of products via www.respect-code.org. On the platform, every detail about the spinning, knitting, dyeing, printing, and confection of the traced products are only a click away, with accessible, summarized, and detailed data about the product, its social and environmental performances, the brand and its values, the supply chain and its mapping. The responsibility relies on the firms providing the data. Independent verification should prevent against any kind of misuse.

The implementation is simple. During the confection process, a Respect-code—materialized by a digit code or a QR-code, is sewn or printed on the garment. By entering the digit code or by scanning the QR-code on any device, the consumer is then able to access, before or after purchase, all details about this particular product, whether in the store, directly on the e-shop or at home.

A world map shows the path that the product has followed all the way from raw material to the retailer. All the different production steps are detailed, mentioning the different factories involved, including addresses, pictures, and contact details. Audits and certificates obtained by the factories forming part of the supply chain can also be viewed. The platform also publishes product-linked indicators, such as CO_2 emissions, water-use footprint, and product-linked certifications.

Additional information can be found about the batch, i.e. the number of items of the same product ordered by the brand on a certain date from one of its direct suppliers. This includes the order quantity, the shipping date, and the person in charge of that particular production batch. Upon demand from the traced company, for competition reasons, some data could be kept internal and not be shown on the public platform.

An example of a traced production can be seen by simply scanning this QR-code (Fig. 1):

Fig. 1 Example of a QR-Code from Dibella GmbH, an organic cotton napkins distributor

2 The Added Value of Traceability

Thanks to traceability, any consumer may have access to each and every step of the supply chain of the product he or she has bought or is considering buying, with full awareness of the social and environmental impacts this choice implies. This is not just a marketing argument; it is about setting standards of best practice, in that a product's provenance, history and environmental impact are openly available to the general public. Brands, as we know, are eager to present a favorable, ethical image, to enhance their reputation and to promote their likeability among existing and new customers. Factual information such as minimum wage, water footprint, CO_2 emissions, various existing certificates and audits are verified by third parties. This reinforces a brand's reputation, shows its responsibility, and increases its credibility towards consumers.

Having developed and practiced traceability for over a decade, we have, nevertheless, noticed that it is often perceived as somewhat of a risk by companies, reluctant to allow their competitors access to their suppliers, to divulge details of manufacturing, or to disclose production data which could potentially be in conflict with the image that their marketing strategies hope to convey. I remain convinced that when well-managed, the spirit of transparency is an extraordinarily efficient tool, creating huge advances in information management, setting new values of excellence and providing companies with the opportunity to share their daily life with the outside world.

Traceability leading to transparency will automatically bring the production of goods to higher CSR standards: better control and increased efficiency in the production lines and the potential acquisition of new customers. The more data is provided, the more stringent will be the analysis process and the optimization potentials. As an example, in the dyeing production step, the choice of a dyeing unit could be made according to the one using less water in its process.

Being transparent about your supply chains re-enforces the confidence of all stakeholders and consumers in the brand and, additionally, boosts your staff's motivation by sharing one, overriding vision within the company.

Opacity in both production and business activities is harmful to all concerned, but, paradoxically, it is those companies who have not yet understood that the world has changed who are most at risk. One should always bear in mind that a customer can become the best ambassador of a brand. On the other hand, tomorrow's clients may well sanction the brand if they feel that they have been deceived or disappointed.

In most cases, a branded product is retailed without any information other than the brand, the selling price, the materials it is composed of and a "made in" indicating origin, but even this is not always the case. This is the angle of vision of a today's consumer, represented as an "eye" on the right side of the chart below. At the bottom, the second "eye" provides full information for each step of the supply chain. This vision, seen from multiple angles, is made possible by traceability (Fig. 2).

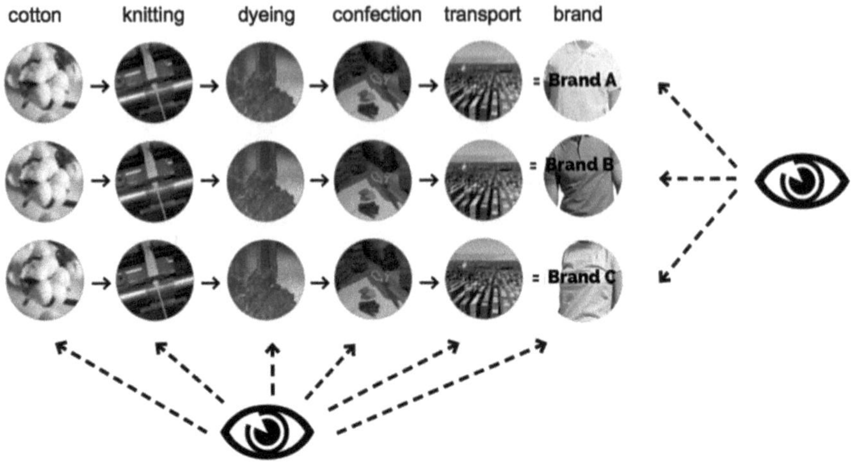

Fig. 2 The two visions of a consumer about a textile product

Better-informed customers make better purchase decisions, which are in line with their values and correspond to a more holistic vision of consumption. In times of economic or political uncertainty, we cannot underestimate the need for clear, factual information in order to make informed choices. Companies running their businesses in a transparent way will always be a step ahead as Dibella GmbH having chosen a complete transparent approach, positions itself as responsible and innovative.

3 The Unlimited Power of the Consumer

According to Kotler and Armstrong (2015), over the past 60 years, marketing has moved from being product-centric to being consumer-centric. Today we see marketing as transforming once again in response to the new dynamics in the environment. We see companies expanding their focus from products to consumers to humankind issues. Nowadays companies shift from consumer-centricity to human-centricity and profitability is balanced with corporate responsibility.

I believe that it is essential to move away from the individualism predominant in the private sector to a more collective approach. A firm develops a more caring and collective vision by inviting all kinds of stakeholders, not only customers and suppliers, but also trade unions, NGOs, and certification bodies. Being focused on the common interest demonstrates a sense of responsibility for businesses and consumers alike.

The political influence linked to the act of buying, which only the consumer possesses, is obvious, see the #whomademyclothes initiative described below or the new trends in the food sector (organic, vegan, locally produced food). It is important to sensitize the consumer to his/her own competence, so that he/she may consciously use this information in supporting a worthwhile cause. The issue here is that this kind of education invokes what could be defined as the syndrome of the elector who does not perceive the importance of his/her participation and trivializes his/her single vote.

Awareness strengthens individual motivation to take position, as when a political party gives voting recommendations to its members. An example of this is the "Who made my clothes?" campaign led by Fashion Revolution, a not for profit Community Interest Company (UK) founded in 2013, after the Rana Plaza complex collapsed in Dhaka. This initiative has encouraged consumers to use their own power, tackling the brands directly on their social networks. By asking the simple question #whomademyclothes to brands, they have ignited a global conversation about supply chain transparency, and started to inspire people to think differently about what they wear. In 3 years, according to Fashion Revolution, the movement has grown to span 92 countries and to reach millions of people (Fashion Revolution 2017).

In the future, we could even imagine that some form of public aid or tax incentives could be granted in the textile sector to all companies providing full access and visibility on their production. Similar incentives are already in place when investing in home thermal insulation in Germany (BMWi 2017) or for the purchase of an electric powered car in France (Ministerè de la Transition Écologique et Solidaire 2017). Such incentives would encourage companies to be fully traceable and transparent and thereby increase benefits for all.

4 Traceability, a True Framework for Transparency

Launched in July 2011, the Greenpeace Detox campaign has since mobilized hundreds of thousands of people around the world to challenge major clothing brands to eliminate all releases of hazardous chemicals from their supply chains and products. So far, the campaign has been able to secure public commitment from nineteen international fashion companies (Greenpeace 2011).

But to go beyond isolated actions, a specific framework should be settled with a list of metrics including chemical uses, water footprint, and CO_2 emissions, verified by third parties such as certification bodies and independent initiatives that work on improving conditions for workers. Consequently, companies, their products, and services will be increasingly differentiated through a combination of "soft values"—social and ecological achievements—and "hard facts" such as the raw materials involved and the origin of the product. The aim of tracking a wide range of information within a supply chain is to aggregate and convert the data into clear and structured messages to all stakeholders and to the outside world including regulators, investors and end consumers. The challenge for companies is to make this mass of information available in a transparent way. As a result, from a state of semi-ignorance about the products available on the market, the responsible end consumer will be able to buy with full knowledge of what his/her act of purchase implies on a global level. This is respect-code.org platform's main contribution.

Taking an example from the food sector: fruit or vegetables bought out of season, transported over thousands of miles should clearly indicate their contribution to the emission of CO_2. It should be made clearer to the customer when he/she purchases a product whether it comes from a far-off country, necessitating transport, or is locally produced, corresponding to its seasonal cycle. I feel that it is important in such cases that we create awareness in a sensitive and respectful way and do not impose moralistic dictates. If the well-informed consumer decides to buy cherries from Chile in the midst of winter, then that is his/her choice. But is it the responsibility of the distributors to offer cherries in mid-winter instead of simply waiting for summer when they are locally available? I feel it is too easy in such cases to place the entire blame on the consumer.

In the same manner, it is also of capital importance to give full details regarding the social conditions under which a garment is produced. This does not mean judging commercial decisions made by the producers, but rather presenting factual details to the buyer, thereby allowing him to make his/her choice with full knowledge of his/her contribution. Excuses based on ignorance—"I did not know"—would then belong to the past. Today end consumers should be able to affirm: "We want to know as much as we can about what we are buying".

In order to render every action possible with the intention of achieving the lever effect, and to provide transparency on all products, a total simplification of information is necessary. To summarize: one needs some clear indicators accessible by all. These could be based on a model, e.g. "green–orange–red light" that the consumer can immediately integrate. In the same way as the consumer analyses the price or expiry date on food products, this reflex should be widened to a broader range of indicators such as the energetic cost for our planet or the social and ecological conditions under which production takes place.

5 The Product Balance Sheet (PBS)

The Product Balance Sheet (PBS), an indicator developed by our company, provides a traceability level and data related to a product in the form of a global assessment, similar to the balance sheet of a company, and rates it on a numerical scale or according to a code of colours ranging from green to red. The data includes hard facts: the description of the product, pictures, selling price, composition, place of production, as well as soft values: audits and certifications related to social criteria such as working conditions, health and safety requirements, and environmental criteria such as the water-use footprint and CO_2 emissions.

By bringing together the different criteria in the form of an index, one creates a complete vision of a product and helps the consumer in his/her purchasing decision, as long as the information remains simple and easily understandable. It will increase the value of a product in the consumer's mind. Within a traceability approach, the implementation of the PBS requires little work, as our traceability platform respect-code.org already integrates all of the necessary data.

Some might want to go further and analyse the whole life cycle of a product (LCA). The LCA takes into consideration the environmental impacts of a product by considering the major stages of a product's life beginning with raw material acquisition, which includes material harvesting, processing, and transportation to manufacturing sites. The manufacturing stage includes product manufacturing and assembly, packaging, and transportation for final distribution. The analysis is also about energy consumption, CO_2 emissions, water-use footprint during normal product life, possibly the product reuse (refurbishing, material reuse) and finally waste management/end of life, including recycling.

6 Accountability, Verification and Independent Monitoring

Accountability requires that each stakeholder must be committed to providing all necessary information concerning his/her area of activity. In a first step, all the information collected will be put online for internal use exclusively. Thereafter, it is up to the company to decide which content will be communicated externally. The cost of producing this information will be partially born by the company, as each stakeholder of the supply chain will produce information related to its activity. The main principle here could be described as 'everyone putting their own house in order.' Internal transparency seems to me to be essential to the good practice and smooth running of a company.

Confidence is healthy but control is necessary. Therefore, an on-going process will manage all kinds of information provided by the stakeholders, verified by third parties such as certification bodies and independent initiatives, before they are put online.

For obvious reasons of credibility, and to avoid playing the roles of judge and jury simultaneously, independent monitoring is compulsory. For this, collaboration and coordination of various organizations are essential, among them independent initiatives like Social Accountability International (SAI), of which I was a member of the advisory board during a decade, Business Social Compliance Initiative (BSCI), or Fair Wear Foundation (FWF), or labels like Oeko-tex, GOTS, Max Havelaar Fair Trade, EU Ecolabel and many more. Finally, the collected information will be posted online and updated continuously.

There is currently no real incentive to improve working conditions or the use of more environmentally friendly materials. This step, which is more than just an economic move, must go through the channels of legal institutions, via the public sector once again. If it were up to me, non-identifiable products, according to certain well-defined international criteria, would no longer be sold; just as a car without a license plate is not allowed on the roads. It is crucial that end users should not only be able to compare prices and product attributes but also that they be aware of the history behind the products they may or may not purchase.

When the pressure for more transparency put on global players and medium-sized companies by all kind of organizations and the public eye is not effective, the situation requires the intervention of the public sector. The case with the Modern Slavery Act voted by the Parliament of the United Kingdom is worthy of attention: since March 2016, companies doing business in the UK with an annual turnover of over £36 million have a legal obligation to declare what steps they have taken to ensure that slavery and human trafficking are not taking place within their business or supply chains (Modern Slavery Act 2015 2014).

7 Trust Really Matters

Expressing companies' values in a transparent way to the external world has a great and positive impact. Trust really matters, but how can we earn it? Laying all cards on the table in full view of everyone will create a feeling of trust among customers and increase credibility, a valuable asset that costly marketing campaigns simply cannot fabricate. Sustainability is now present in most main textile brands. Independent initiatives promote fair trade and working conditions, while ecological impacts are being monitored by audits. There have been considerable advances over the last decades. But if we want to learn more about what we eat, why should we not also inquire about the garments that we wear, as well as any other consumer goods? The sustainable soft values of each product should always be mentioned so that the consumer can determine his/her buying priorities at first sight. Only a real traceable on-line tool was missing, but today this tool exists.

8 Conclusion

The contribution of traceability, highlighting soft values as well as the hard facts of a product, is immense. If until today hard facts are the criteria to differentiate one product from another, tomorrow the soft values of a product will also count, and more importantly than ever before. The respect-code.org platform enables consumers to make conscious choices in the market place leading to purchasing decisions in retail, wholesale, and e-commerce settings. The environmental and social impacts of each production step are revealed to all stakeholders and will help in ensuring that workers' rights are respected. The way in which those soft values are published has to be clear and user-friendly, enabling the consumer to compare and weigh them with ease. This will allow him/her to make his/her purchasing decision faster and more easily and to assume greater responsibility for his/her choices. The aim is not to name and shame, but rather to appeal to everyone's common sense, as, obviously, industry will no longer be able to operate if resources are in short supply and if the environment is in danger.

References

BMWi (2017) Ready for the next phase of the energy transition. http://bmwi.de/EN/Topics/Energy/Builings/energy-effficiency-strategy-for-buildings.html. Accessed 15 Jun 2017

Fashion Revolution (2017) http://fashionrevolution.org. Accessed 15 Jun 2017

Greenpeace (2011) The Detox Campaign. http://www.greenpeace.org/international/en/campaigns/detox/water/detox/intro/. Accessed 5 Jun 2017

Kotler P, Armstrong G (2015) Principles of marketing, 16th edn. Prentice Hall, New Jersey

Ministerè de la Transition Écologique et Solidaire (2017) Bonus-malus écologique, prime à la conversion et bonus vélo. http://www.developpement-durable.gouv.fr/Bonus-Malus-definitions-et-baremes.html. Accessed 15 Jun 2017

Modern Slavery Act 2015 (2014) This page contains an overview of the modern slavery act and supporting documents. https://www.gov.uk/government/collections/modern-slavery-bill. Accessed 28 Aug 2017

Respect Code (2017) www.respect-code.org. Accessed 15 Jun 2017

Sustainable Fashion: From Production to Alternative Consumption

Samira Iran

1 Introduction

New garment sales have increased by 60% in a period of 10 years in the UK (NIRI 2006). Meanwhile, textile waste in the UK is recognized as the fastest growing fraction of household waste (DEFRA 2008) and the global textile industry has serious negative environmental effects within its supply chains (BSR 2009). The unsustainable characteristic of the fashion industry and the significant effect of fashion and clothing on global sustainability development (Allwood et al. 2006) have stressed the need for fashion consumption sustainability and an increased demand for eco-friendly clothing (Williams et al. 2005) has started to influence the fashion industry (Nagurney and Yu 2012). Consequently, the focus of sustainability research has been shifted towards the consumption phase of the fashion supply chain (e.g. Fineman 2001; Birtwistle and Moore 2007; Carter and Rogers 2008; Niinimäki 2010; Armstrong et al. 2015).

Against this background, the aim of this chapter is to illustrate the concept of sustainable fashion and to answer several fundamental questions regarding sustainable fashion: What does sustainable fashion mean and what are its elements? Who are the actors in this industry? How important is the consumption phase of sustainable fashion? What are alternative fashion consumption models?

In this paper, firstly, the concept of sustainable fashion is defined and elaborated in the context of the three categories of governance, production, and consumption and key drivers and actors of each phase are discussed. Due to the importance of the consumers' role, the consumption phase is specifically emphasized. Secondly, innovative and alternative fashion consumption possibilities are introduced described by term "collaborative fashion consumption" (CFC).

S. Iran (✉)
Institute of Vocational Education and Work Studies, Technische Universität Berlin, Berlin, Germany
e-mail: s.iran@campus.tu-berlin.de

S. M. Jastram, A.-M. Schneider (eds.), *Sustainable Fashion*, Management for Professionals, https://doi.org/10.1007/978-3-319-74367-7_13

2 Sustainability in Fashion

Fashion (in particular garment fashion) and clothing are two different concepts. While clothing is regarded as one of the basic needs of humankind, fashion is based on desire instead of need. But it seems that this nuance is not apparent in most contemporary societies, where the majority of the population have more clothes than they really need or could ever possibly wear (Black 2008). Therefore, in this chapter the terms fashion and clothing will be applied interchangeably.

The fashion industry is faced with manifold environmental, social, and economic issues because of the existing fast-paced fashion trends. Water and air pollution caused by dying processes, pesticides that are used to grow the raw materials, artificial materials that are used in the production of clothes, and CO_2 emissions caused by lengthy transportation routes are only some examples of the environmental problems caused by the unsustainable fashion industry. For instance, conventionally grown cotton, as one of the most popular clothing fibers, is one of the most water—and pesticide—dependent crops (Claudio 2007).

Moreover, as published in the media, workers in developing countries are suffering from working conditions that fast fashion companies put them through (e.g. Donaldson 2016). As an example, one can refer to the collapse of Rana Plaza in 2013, in which thousands of workers died or were injured. Additionally, child labor (e.g. Moulds), low wages (e.g. ILO 2014), and the lack of insurance could be listed as few social complications that have been seen in the unsustainable fashion industry. Considering the manifold problems that have arisen in the last decades, the concept of sustainable fashion consumption has been recently surveyed by various researchers (e.g. Fletcher 2008; Black and Anderson 2010; Schultz 2010; Armstrong et al. 2015).

3 Sustainable Fashion Discourse: Definition and the Developments

After the "Hippie" revolution in the mid-1970s, a second phase of fashion evolution evolved in the 1990s with a focus on sustainability issues (Black 2008). Esprit presented its first Ecollection in 1994. In 2007, the first environmental strategy was written for Marks and Spencer, which helped the agenda for others to follow (Black 2008). Hamnett in collaboration with the United Nations, developed the first certificate for sustainable clothing (Black 2008).

Joergens (2006) describes ethical fashion as fashion clothing that is manufactured considering fair-trade principles and environmental standards of the production process. However, there is a fundamental debate on the paradoxical nature of sustainable fashion. On the one hand, production efficiency has increased affordable fashion availability (Eder-Hansen et al. 2012). On the other hand, sustainability demands durability and the practice of recycling (Morgan and Birtwistle 2009).

There are trade-offs to be made in all phases of clothing production and consumption considering available materials, costs, and time constraints. Each of these decisions has an impact on the next phase/s. For example, the chosen fiber will impact the whole product life cycle- from cultivation, production, manufacturing, distribution, consumer laundering, to reuse and final disposal (Fletcher 2008).

Therefore, in our more comprehensive definition, we refer to sustainable fashion as:

> Clothing that is designed, produced, (re-)used and disposed in a way that is aligned with the concept of sustainable development. Thus, we consider not only the sustainability of the production of the garments but also the sustainability of their usage and post-usage phase. The goal of sustainable fashion is thus to reduce environmental and social impacts of the conventional fashion industry, whose production processes are historically amongst the worst polluting, as well as being associated with systematic labor- and human rights law abuses in manufacturing countries. Each step in the complex production and usage cycle of fashion needs to be transformed in order to create sustainable fashion (Stanszus and Iran 2015: 154).

In sustainable fashion, materials are initially carefully selected among those that have the least possible environmental impact. In the next step, garments are designed in a way to not only minimize the waste of cutting fiber or yarn but also to make intelligent fashion items that have timeless designs and are long lasting (Stanszus and Iran 2015). The processing of the garments becomes more sustainable. For instance, natural dyes will be used. Less transportation will be planned and the production moves toward local manufacturing. "In sustainable fashion, networks of trade unions, governmental institutions, and corporations are managing workers' rights to ensure fair treatment" (Stanszus and Iran 2015: 154). In the consumption phase, consumers will wash their garments less frequently and at a lower temperature, they buy less clothing and as soon as they do not want to continue using the clothes, they will find a way to recycle, reuse or redesign the garments (Allwood et al. 2006). "Sustainable fashion is nothing which is ready for sale; it needs to be created by joint responsibility of companies, consumers, and others" (Stanszus and Iran 2015: 515).

Kate Fletcher considers three phases for innovation in the context of sustainable fashion as driven by legislation, technology-based innovation, and innovation of the consumption phase (Fletcher 2008). Being inspired by this categorization, drivers of sustainable fashion will be studied here based on the three key categories of governance (e.g. governments, NGOs), production (e.g. manufacturers, suppliers, designers), and consumption (consumers), where governance can influence both production and consumption. However, drivers influencing sustainability of the fashion industry are not limited to the ones presented in Fig. 1.

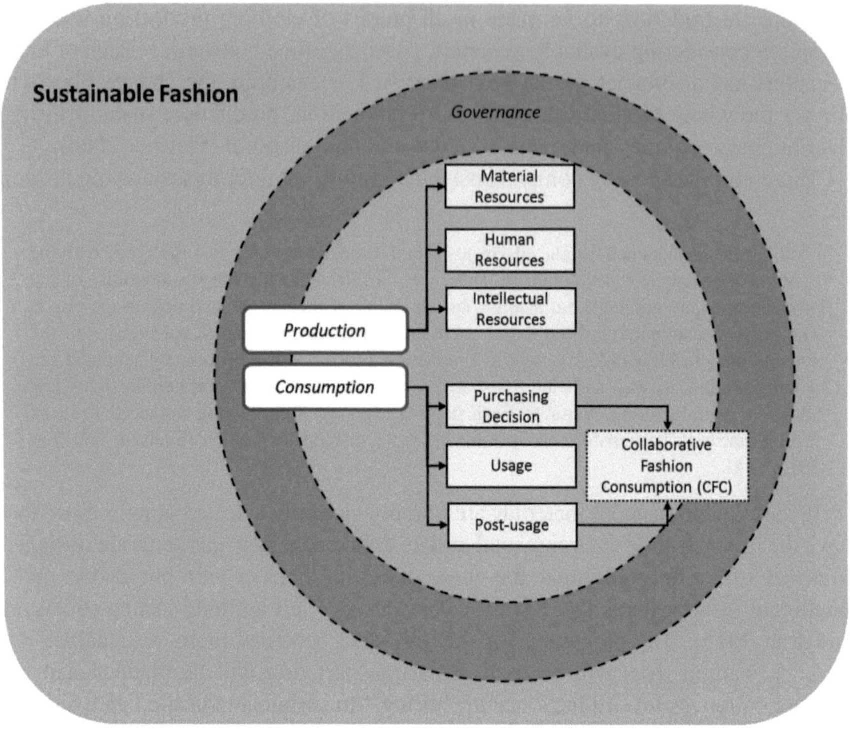

Fig. 1 Drivers of sustainable fashion

4 Sustainable Fashion Governance

"Legislation has had a formative influence on sustainability innovation in the fashion and textile sector to date" (Fletcher 2008: 54). Some standards and regulations are currently generated for production of fashion items in different countries across the world. However, it seems that an effective enforcement mechanism toward sustainable fashion is still missing in the textile sector, as exploitation of workers and environmental problems of the unsustainable fashion industry are still issues to be solved. From a sustainability perspective, it is desirable that fashion items are being produced in a sustainable manner and follow the guidelines of sustainability such as ISO 14000 (Lo et al. 2012). Moreover, fashion companies can apply their positive image in terms of sustainability as an effective marketing strategy (Nagurney and Yu 2012). Legal boundaries in combination with consumers, NGOs, and media pressure can control production in a market-based economy. One of the areas that governments have globally focused on is waste management, where the policies emphasize restrictions on landfill, treatment, and disposal of hazardous waste, and increase the practice of recycling (Morgan and Birtwistle 2009). Moreover, tight import/export regulations as well as strict labeling are some of the strategies that governments could apply in order to promote and control sustainability in the fashion market. With a set of import/export rules, government could support and promote local production and as a result decrease the carbon emission that results from transportation in a global textile supply chain.

Furthermore, there is an emerging trend of alternative fashion initiatives. Common regulations that are used for traditional and conventional fashion businesses cannot directly be applied to the alternative fashion initiatives or are sometimes not as effective as needed. For instance, in the case of clothing swapping, the classic buyer and seller cannot clearly be defined and distinguished or in case of online second hand buying, the pricing of garments cannot follow the pricing strategies that are used for the new fashion items. In addition, in all of the online alternative fashion consumption models a mutual trust is required (Gata 2015). Hence, setting specific laws and regulations for the alternative businesses could protect the rights of both the businesses as well as users/consumers.

The important role of NGOs in pushing and forcing companies to respect the rights of workers cannot be neglected. In the 1990s, NGOs exposed the labor abuse in the supply chain of brands, such as Nike or Gap, which resulted in the introduction of codes for minimum levels of worker rights as well as some legislation-driven changes (Fletcher 2008). One example of NGOs that are focusing on clothing production is the Fair Wear Foundation (FWF). FWF is an independent, non-profit organization working with companies and factories to improve labor conditions for garment workers. Another example is the Clean Clothes Campaign (CCC), whose offices are widely spread around the world. The CCC is dedicated to improving working conditions and supporting the empowerment of workers in the global garment and sportswear industry. Since 1989, this campaign helps to ensure the respect for fundamental worker's' rights. Educating and mobilizing consumers, lobbying companies and governments and offering direct solidarity support to workers as they fight for their rights and demand better working conditions are some working areas of this NGO.

Sustainability can be applied as a framework to reach environmental and socially conscious consumers and at the same time to increase the overall brand image of the company in developed countries (Faisal 2010). Corporate Social Responsibility (CSR) as a prominent phenomenon in textile and clothing, involves the advancement of internal corporate programme and systems, social auditing and certification, as well as the integration of companies in multi-stakeholder initiatives (ILO 2005). The government can positively increase sustainability of the fashion industry by emphasizing the importance of the CSR activities of companies as well as having control over the performance of the garment industry. The governments have generally played the key role in altering the conventional fashion industry toward a more sustainable one. Legislation and standards should be considered in all stages of the textile processing chain.

5 Sustainable Fashion Production

Some researchers have strongly criticized the apparel industry for their unsatisfying efforts to produce an ethical supply chain (Bendell and Kleanthous 2007). However, increasing demand for eco-friendly apparel has impacted clothing production and forced companies to offer green items by e.g. using sustainable materials to satisfy the green consumers' demand (Sampson 2009). For instance, H&M is claiming to be one of the leading users of organic cotton in the world. Additionally, in Europe, high street retailers, such as H&M, have launched Eco-collections, and in the US, Gap Inc., has focused its efforts on decreasing the environmental impacts of its supply chain (Jegethesan et al. 2012). Beside such attempts, many researchers and experts have investigated the production phase in order to achieve sustainability of fashion products. Hence, some innovative solutions have been generated during the last decades. Endeavors of these production related-innovations have mostly been towards achieving more efficiency in one or more resources. One can categorize their efforts into three sub-groups, based on the kind of resource(s) they focus on: material resources, intellectual resources and human resources.

5.1 Material Resources

The kind and the amount of materials that are used in the production process of sustainable fashion items should be carefully decided. As commonly understood, sustainable fashion means using natural materials like cotton in garments instead of using synthetic, oil-based materials like nylon. However, according to the investigations of ecological impact and ethical issues regarding production of both mentioned materials, it is difficult to say which one is the better choice in terms of being eco-friendly. Each of these materials has its own sustainability issues that cannot be neglected. For instance, production of manufactured materials generally consumes less water, however, polyester and acrylic have high emissions to air (Fletcher 2008). The majority of problems with cotton occurred in Africa and India, where small-holding farmers in order to increase yields, fulfill the demand and maximize the profit by using pesticides that cause illnesses and death of workers. About 117,000 metric tons of organic cotton were globally produced during the 2013–2014 (OTA 2015), however, even for these organic cottons a range of chemical substances is applied in the production process, which questions the sustainability of using natural fabrics instead of manufactured fabrics.

Furthermore, more attention is currently given to the recycling of textiles back into fibers. Patagonia was a pioneer in the field when it began selling its "post-consumer recycled" line in 1993. Using plastic bottles to produce garments is now practiced by other companies such as The North Face. However, the amount of recycled textiles is still inconvenient. For instance, in 2014 the recycled rubber, leather, and textiles were about 9% of all recycled items in the USA (US EPA 2014).

In Germany, Anke Domaske, the inventor of "QMilch" introduced a fabric made entirely of spoiled milk. However, using such innovative/more sustainable fabrics for garment production on a bigger scale might not yet be possible. In summary, there is a need to extent the amount of organic material and the smart usage of manufactured ones, as well as to develop an efficient technology for recycling material.

For many manufacturers, the improvement of production technology is a true way of being innovative. During the production processes (e.g. dyeing, drying, and finishing), chemical products and natural resources are intensively applied and are generating a high environmental impact (Rinaldi and Testa 2015). Although using eco-friendly technologies in all the clothing-chain helps reduce the negative effects of the garment industry, this is still not a real answer to the problem; since relying on technology to fix all of the problems just decreases the needed consideration for the soft cultural change (Fletcher 2008).

Due to the globalization process in the last decades, most of local productions are now shifted to different countries where material and labor forces are cheaper. In these countries, CSR is often less prevalent, NGOs are less powerful and consumer's decision-making is more based on their basic economic needs (Jastram 2007). Consequently, different countries often located far apart are engaged in the production process of garments. Therefore, massive amounts of energy and resources are consumed merely for transportation of materials and manufactured goods. For instance, in 2001, Fran Abrams and James Athill tracked a pair Lee Cooper jeans sold in a high street store in the UK on its global journey of 40,000 miles by land and sea, from producing the fiber in Benin to manufacturing in Tunisia (Guardian 2001).

5.2 *Intellectual Resources*

Designers are responsible for the different steps of their decision-making procedure, as their decisions consequently have impacts on environmental and social aspects of the whole product life cycle and sustainable consumption patterns (Tischner and Charter 2001). They choose fabrics and use them in their collections considering their design, cost, availability, and time constraints. The main challenges at this level are the selection of natural or manufactured fibers (e.g. Recolution uses organic cotton and GOTS certified materials in its collections), decisions for new or recycled fibers, design or re-design (e.g. Bag To Life works with the concept of upcycling and re-design), using mono or multi fibers (e.g. Patagonia works directly on the fabrics that are used in its collections), design for short or long term (e.g. Uniwearsal claims that they produce timeless designed apparel), or trust in suppliers or direct control of the material supply chain (e.g. Gossypium produces sportswear in its own workshops to have direct control over the production).

One of the most outstanding responsibilities of designers is to design a charming sustainable item, which satisfies the needs and desire of the consumers (Zafarmand et al. 2003). Often, sustainable consumers have difficulties finding suitable sustainable clothes because green clothes are generally noticed as shapeless, colorless, overpriced (Meyer 2001). Hence, the aesthetic values need more consideration when designing sustainable clothes (Fletcher 2008). The point is that the desire for novelty will not and should not be eliminated, however development of alternatives is needed (Black 2008). "Eco-chic must aspire to being the norm not the exception" (Black 2008: 18).

Slow fashion and Cradle to Cradle are two recent concepts in the design phase approaching sustainable fashion. Cradle-to-Cradle as a design concept was initiated by William McDonough and Michael Braungart in 2002. This concept stresses the need for increasing the efficiency of production techniques and achieving waste free production. It should be possible to recycle, reuse, or re-consume all materials and to bring them back to a new life cycle (McDonough and Braungart 2010; Butler 2007).

Slow fashion emerged as a reaction to the rapid trends of fashion buying behavior or fast fashion aiming at decelerating the fast pace of fashion. Slow fashion is defined as clothing that starts with careful choices and well-paid individuals and then remains to be used for years (Sayer 2007). Designers are encouraged to integrate new insights into their patterns to make long lasting designs and thereby slow down the fashion pace. The concept of "everybody designs" presented by EzioManzini is another example of innovative thinking in the fashion industry.

Integrating the end-users into the design process could increase the usage lifetime of the products. Moreover, some companies have realized the benefits of involving users in the development and diffusion of innovations (Baldwin and von Hippel 2011; Füller et al. 2012). Concepts like prosumption, in which both production and consumption are involved, integrate consumers' ideas into the design process (Ritzer and Jurgenson 2010). Engaging consumers in a participatory design process can also be seen as a way to achieve more sustainability in the fashion industry. Creating social values through meaningful experiences and collaboration can motivate consumers to value their clothes more as well as integrate their tastes into the design and as a result extend the usage phase of the product life cycle.

5.3 Human Resources

"Who should make the clothes and how should these workers be treated" are the main questions, when talking about human resources within the sustainable fashion production chain. Directly after the garment is designed, the first decision to make is whether local workers are hired to make the products or to outsource the production to other countries e.g. Asian, African, or Indian garment manufacturing corporates. Due to the availability of cheap labor, the garment production has recently been shifted to developing countries. This outsourcing mostly causes long transportations as well as labor conditions that are not aligned with the fair-trade rules. The term "Fair-trade" mostly covers the activities and implications of advocating for fair working conditions in manufacturing industries. Some researchers in the field of environmental clothing have paid attention to labor exploitation among other ethical issues (e.g. Dickson 2001; Rudell 2006; Shaw et al. 2006).

The FWF is one of the NGOs that aims at developing fair-trade production in the garment and textile industry. The FWF code of labor practices contains eight labor standards that are made up based on the core labor standards of ILO. In FWF labor standard's code: employment is freely chosen, there is no discrimination in employment, no exploitation of child labor, freedom of association and the right to collective bargaining, payment of a living wage, no excessive working hours, safe and healthy working conditions, and legally-binding employment relationships. These standards should be carefully followed by the manufacturers who are willing to produce sustainable fashion products.

Since the 1990s, the media and public have put pressure on fashion manufacturers to generate legal documents to protect labor rights in garment industries. However, such efforts have not yet completely resolved the problems in the textile industry (see Donaldson 2016). Selecting the human resource in addition to setting employment rules based on fair-trade standards is a key factor in reducing the social problems of unsustainable fashion production.

6 Sustainable Fashion Consumption

People are becoming increasingly interested in the environmental and social consequences associated with their consumption patterns (e.g. Auger et al. 2010; Harrison et al. 2005). This development triggers a growing demand for ethically and sustainably manufactured products and services (e.g. Williams et al. 2005). Nevertheless, companies have been criticized for not taking useful strategies into account despite the fact that the emergence of the sustainability concept, the importance considering the role of the consumers, and changing the lifestyle pattern have lately been emphasized (e.g. Schaefer and Crane 2005; Belz 2006; Thøgersen and Crompton 2009). Schrader explained it in two ways: first, by highlighting the impact of consumption on ecological and social lifestyle of people, and second, by marking the consumers' role as a potential influence on the companies' behavior (Schrader 2007). Therefore, the role of the consumer in reaching better sustainability is noticeable, and consumers are now looking for more sustainable solutions.

In a report by the University of Cambridge (Allwood et al. 2006), a cotton T-shirt, a viscose blouse, and a nylon carpet were compared to each other with regard to their environmental, economic, and social data in various scenarios. The results revealed that shifting the production of the T-shirt and the viscose blouse to the UK saves modestly, since the used energy for transportation is relatively small in comparison to the washing of clothes. In contrast, a change in consumer behavior has a large positive environmental impact (Allwood et al. 2006). It can embrace a large number of reuse, repair, and recycling activities to augment the life-cycle of the clothes (Allwood et al. 2006). In addition to the positive effect of sustainable fashion consumption behavior on the environment, Paulins and Hillery found ethical fashion to fulfill consumers' specifics as well as basic psychological needs (2009).

Kate Fletcher discusses recent sustainable fashion innovations regarding process, product, and consumer focus. She believes that the biggest benefits come from consumers (Fletcher 2008). But it should not be forgotten that fashion consumers differ from other consumers of other product types with regard to their ethical consumption decisions (Niinimäki 2010). For instance, regarding food consumption, consumers show more commitment to sustainable consumption as food has a direct impact on their individual health (Joergens 2006; Ochoa 2010), however, less commitment has been expressed in the fashion sector due to the wrongly perceived indirect effects of the fashion products on their health (Joergens 2006).

There is still an attitude-behavior gap in sustainable clothing consumption. Consumers should become aware of the environmental and social impact of their clothing consumption as well as the alternatives to unsustainable fashion consumption. They should be encouraged to practice sustainable fashion consumption in their everyday life. Many lines of research consider the problem of unsustainable fashion consumption (e.g. Niinimäki 2010; Ochoa 2010; Lewis and Gertsakis 2001; Palmer and Clark 2005). Consumer behavior involves pre-purchase, purchase, and post-purchase components. Some researchers have limited sustainable consumption to sustainable buying behavior (e.g. Balderjahn et al. 2013). However in this chapter sustainable fashion consumption refers to all three phases of purchasing, usage and post-usage of the garments.

6.1 *Purchasing Decision*

Consumers are increasingly taking the ethical attributes of clothing into account (Kim and Damhorst 1998; Dickson 2001). In the purchasing decision step, the significant discussion is to determine the consumer intention towards the fashion products—whether they are willing to buy new or used products. In this regard, the perceptions and attitudes of the consumer play a significant role in addition to the cultural and social values of the whole society. Moreover, concepts, such as sharing, renting, and swapping, can be effective and applicable to decreasing ecological footprints only when people accept to buy the service instead of the product. It is also crucial that consumers understand and practice the sufficiency concept in the decision-making process and avoid over-consumption of material by following the fast fashion trends. Buying sustainable products will not be a complete solution unless consumers combine it with buying less, using the products longer and more efficiently and eventually bringing them back into the cycle.

6.1.1 Usage

The usage step in the consumer behavior process describes how consumers use and maintain fashion products in a sustainable manner. Laundering is one of the most water and energy consuming parts of this step. However, there is a lack of studies that focus on cleanliness, hygiene, and the environmental consequences of laundering (Fletcher 2008). Moreover, ironing and washing temperatures have a significant impact on the energy consumed in the usage phase (Niinimäki 2009). In 2001, Lewis and Gertsakis estimated the possible impact of consumer care to be about 75–80% (depending on the material used in the shirt) of the total ecological impact of a cotton shirt (Lewis and Gertsakis 2001).

In the UK, it is estimated that between 0.8 and 1 million tons of apparel are sent to a landfill yearly. Among them, a minimum of 151,300 tons of clothing and footwear could be directly reused (Bartlett et al. 2013). This shows that consumers sometimes sorted their clothes out, although the garments were still usable and before the garments reached their end of life cycle. These fashion products could be seen as unused potentials and could be used for a longer period of time. Extended usage of the clothes could benefit the environment. It is claimed that if the lifetime of clothes could be increased by 33%, potential footprint reductions and resource cost saving could amount to 27% carbon savings, 33% water savings, 22% waste savings, and 22% resource cost savings (WRAP 2012).

Repairing the clothes could be another way to prolong the clothing lifetime of the garments. According to a study by Greenpeace, 42% of the participants have never repaired their clothes (Greenpeace 2015). Bringing back the trend of valuing and repairing the garments could potentially decrease the unsustainable fashion consumption.

6.1.2 Post-usage

By increasing fashion purchasing, most of the garments are disposed of after being worn only a few times (Birtwistle and Moore 2006). Studies on wasteful consumption that were conducted by The Australia Institute show that about 1.7 billion dollars were spent yearly for clothes and accessories that were not worn at all or only used for a short period of time (Pears 2006). In Germany, each adult between (18–69 years old) has about 18 clothing pieces that have almost never been worn (Greenpeace 2015). This makes for about 1 billion unused clothing pieces stored in German wardrobes (ibit.). It seems that a huge amount of clothes is treated like waste shortly after purchasing. However, in contrast to other types of waste, they stay unused in wardrobes instead of going to a landfill. Analyzing the clothing life cycle reveals the significant role of the consumer in garment waste management (NIRI 2006; Tukker et al. 2008; Birtwistle and Moore 2007). Consequently, some researchers and fashion designers paid attention to this phase of the consumption process to increase the sustainability by re-designing, recycling, and re-using these unused fashion products.

6.1.3 Re-designing

Some small designers now use different methods and techniques, such as upcycling, restyling, reshaping, embellishing, and over-printing, to make new clothes out of the unwanted garments (Fletcher 2008). TRAID is an example of charitable organizations that give unsold charity clothes to the young designer groups to redesign and remodel them into fashionable pieces, which then are sold again. The Upcycling Fashion Store Berlin is also one of the initiatives that sells only garments that have been redesigned and remade out of existing clothes. Different designers who work in cooperation with this store use their innovative design ideas to reshape the existing garments and offer consumers redesigned clothes.

6.1.4 Recycling

The importance of recycling was mentioned in the production phase. However, consumers can play a significant role in this regard. They can find ways to recycle their clothes and use them in other ways. Donating the unwanted clothes is not always the best way, as it has been mentioned, "only about one-fifth of the clothing donated to charities is directly used or sold in their thrift shops" (Claudio 2007: 452). There is a huge load of clothes that are given as donations. For instance, more than 12 million pounds of post-consumer textiles are processed yearly in the Trans-America Trading Company; about 30% of such textiles have industrial use, 25–30% are recycled into fibers to be used as stuffing for upholstery, insulation, and manufacturing paper products, and only 45% of them continue their lives as clothing and even then, not domestically (Claudio 2007). These apparels are imported to the developing countries and are sold for expensive prices, there. Considering the high prices for the local people in such countries as well as the environmental problems caused by transportation of the used clothes to such countries, donating the unwanted clothes is not the best option. One can suggest to recycle and reuse the clothes domestically in order to avoid the consequences of transportation.

6.1.5 Re-using

It is claimed that collecting, processing, and distributing used apparel requires 1.7 kWh of extracted energy per kg of second hand garment recycled (NIRI 2006). However, the energy used to collect, sort, and re-sell used garments requires between 10 and 20 times less energy than for production of new clothes (Laursen et al. 1997). A life cycle assessment study shows that for every kilogram of new cotton clothing replaced by used clothing about 65 kWh are saved, and for every kilogram of new polyester clothing replaced by used clothing, about 90 kWh are saved (Woolridge et al. 2006). In conclusion, it seems that reusing clothes consumes more energy than recycling them. However, in comparison to the energy used during the production of new clothes, re-using saves a considerable amount of energy.

Some unused clothes are passed on to charity shops that sell second hand clothing in local areas. Only 10–20% of them are resold in UK charity shops (Black 2008). Others are sorted, and better quality items are sent to developing countries to be sold in the local market (e.g. Rivoli 2015; Mhango and Niehm 2005; Hawley 2006). There have been some arguments about western clothing exports threatening the local garment industry and culture of the countries that import these clothes. A study conducted by Baden and Barber reveals that the second hand clothing imports are not the only cause for the decline of clothing production and employment in West Africa. Unreliable and expensive infrastructure as well as high costs and lack of available materials and training plus cheap import from Asia are some of the main reasons for this issue (2005). Additionally, second hand clothing trades create substantial employment in countries that are exporting the used clothes as well the countries that are importing those clothes. Promoting the market of used clothes in the same countries, where the used garments are collected might be a better strategy to decrease the environmental impact of the transportation of used garments.

7 Collaborative Fashion Consumption: The Latest Development

One can take a look at the traditional lifestyles and have some idea of how societies used to maintain a harmony between satisfying their needs, and at the same time not depleting the natural resources. One of the common activities in traditional societies was sharing goods and services. These collaborations are nowadays rare due to various economic reasons. Therefore, instead of using existing products, people are often used to enjoying the concept of ownership through buying affordable products. This new ownership concept has led to over-consumption of natural resources, as well as, social and economic injustice.

Before the industrial era, there was no mass production of products. People were consequently valuing their products and were engaging in sharing activities more than what we see nowadays in societies. On the one hand, products are currently cheap and affordable (at least in developed countries) and most of the people can buy their own goods. On the other hand, the lifestyle of people has changed to more individual-oriented, in which the speed of lives is increasing and people are willing to acquire everything by spending the least amount of time possible. As a consequence, the concept of sharing products and services is no longer popular among consumers. However, recently some scholars and practitioners claim that a sharing economy or collaborative consumption has the potential to increase sustainability and reduce the ecological and carbon footprint in a number of key areas (Schor and Fitzmaurice 2015). It is also claimed that collaborative consumption could lead to more social connections between citizens (Botsman and Rogers 2011).

Collaborative consumption refers to "the redistribution of used products in which two or more persons (re)use the same product in a different period of time regardless of whether the ownership is transferred or if a monetary or non-monetary fee has been assessed" (Iran and Schrader 2017: 472). This understanding of collaborative consumption is aligned with the definition introduced by Botsman and Rogers (2011). Information and Communication Technologies (ICT), explicitly Web2.0 technology, offered and extended the sharing possibilities. The previous local redistribution practices can now be promoted on a largerscale and with less transactional costs (esp. information and contracting costs, Schrader 2001). Airbnb, Uber, Drivenow, and Spotify are some well-known examples of the fast growing companies that offer collaborative consumption opportunities. The Internet offers people a higher availability of sharing and collaborative consumption options.

Collaborative consumption has been also practiced in the fashion industry. CFC offers consumers the option of making their used clothes available on the market and/or taking advantage of using the already existing fashion items. "CFC embraces fashion consumption in which consumers, instead of buying new fashion products, have access to already existing garments either through alternative opportunities to acquire individual ownership (gifting, swapping, or second hand) or through usage options for fashion products owned by others (sharing, lending, renting, or leasing)" (Iran and Schrader 2017: 472). CFC is not a new concept and has been practiced in families, between friends, or even in the flea markets. But as mentioned before, nowadays it can be practiced on a broader scale through available online and offline platforms. Some businesses have already recognized the market potential of CFC and offer online or offline platforms, where different types of CFC (e.g. swapping, renting) are offered in exchange for monetary compensation.

In 2011, Botsman and Rogers introduced three models of collaborative consumption based on different forms of transactions. The first collaborative consumption model, Product Service Systems (PSS), can be described as 'pay for the benefit, not the ownership' (Botsman and Rogers 2011). It disrupts the traditional models of individual private ownership and shifts the concept of ownership from material ownership to service ownership (Botsman and Rogers 2011). In the fashion industry this model can be seen in clothing and accessories rental. There are different websites and shops trying to apply this concept to allow people to rent a product and use its service instead of buying it. Fashion Hire, Girl Meets Dress, Kennedy Purple, and One Night Stand are some examples.

Redistribution Markets, the second model of collaborative consumption is defined as exchanges that make second hand goods available where they are needed (Botsman and Rogers 2010). By increasing fashion purchasing, most of the garments are disposed of after being worn only a few times (Birtwistle and Moore 2007). Instead of being stocked in the wardrobes, the massive volume of not-used (or once-used) clothing could be worn by others who need them or can use them again. Cloth swapping is embedded in redistribution markets and has been developed "from private parties among a small group of girlfriends to high fashion, big warehouse ticketed events" (Botsman and Rogers 2011: 90). Big Wardrobe, Covert Candy, SwapStyle, and Fashion Reloaded Clothing Swap are some examples of fashion swapping websites.

The third model, Collaborative Lifestyles encourages "people with similar interests [to] band together" (Botsman and Rogers 2010: 85). Here people can exchange their time, skills, spaces, and so on. The most applicable skill in fashion industries is the ability to design or re-design the garments. Used clothes can be redesigned through methods and techniques such as upcycling, restyling, reshaping, embellishing, and over-printing. This is not limited to charity clothes but can happen among friends and neighbors where people redesign their old clothes and use them again or share them with those who need them. A growing number of sewing cafés are opening in different cities in Germany. People can use the space and available materials in addition to the knowledge that is offered either by organizers or other fellow members who are coming to the sewing café to repair and redesign their clothes. Skill sharing and social connections that are happening in such sewing cafés are some of the important motivators for participation in such platforms (Hirscher and Iran 2017).

There are only a few articles that explicitly focus on collaborative consumption and even fewer focus on CFC. Recently, Schor and Fitzmaurice (2015) proposed a typology for a sharing economy. They introduced four types of collaborative consumption based on two dimensions of "market orientation towards profits and value creation" and "organization nature" (Schor and Fitzmaurice 2015). In 2017, Iran and Schrader offered a typology of CFC and categorized CFC practices into peer-to-peer and business-to-consumer types. "The peer-to-peer type encompasses forms of CFC in which fashion products are passed from one consumer to another. Peer-to-peer could be organized by peers themselves either through online or offline platforms or by a company in a service-based form" (Iran and Schrader 2017: 473). In Peer-to-Peer CFC six different sub-types are introduced as: swapping parties, gifting, sharing, lending, borrowing, and buying as second hand clothing (Iran and Schrader 2017). All of these practices can happen offline and among a group of friends, colleagues, relatives, etc. However, by using social networks such as Facebook, people can widen their network and practice CFC even with those who they don't know or know less. For instance, in Germany, consumers can use networks such as "Klamottentausch" (Clothing Exchange) for clothing swapping and "Kleiderkorb" (Clothing Basket) or "Kleiderkreisel" (Fashion Spinner) for gifting, swapping, or selling second hand clothing. Kleidertausch (clothing exchange) is another Facebook page that announces different clothing swap parties all over the Germany. Companies can facilitate Peer-to-Peer CFC on a largerscale. For instance, Ebay is an online platform that consumers can use to sell or buy second hand clothing. "Bag Borrow or Steal" is an example of a company, where users have the opportunity to sell or rent-out their used luxury bags. In Germany, "Mädchenflohmarkt" (Girls Flea Market) is a website, "where consumers can put their used clothes on the website and earn up to 80% of the sales price" (Iran and Schrader 2017: 474).

The other type of CFC that has been mentioned is business-to-consumer CFC (Iran and Schrader 2017). "The companies offer either services as substitutes for product ownership (renting and leasing) or second hand retail service to make the purchase of new products dispensable" (Iran and Schrader 2017: 474). People have had the possibility of renting special costumes from renting shops. Nowadays there are some clothing libraries that offer subscription-based services (e.g. Pedersen and Netter 2015). For instance, Kleiderei is a clothing library in Germany. Once a month members of this library receive four pieces of clothing for a monthly-fee of 39 €. Apart from fashion libraries, there are different websites for renting fashion products. Kennedy Purple is a company that rents out designer handbags, jewelry, and fashion accessories. Leasing is another sub-type of business-to-consumer CFC that has been less practiced. In Germany, "Lütteleihen" (Baby Clothes Rental Service) offers leasing for baby clothing. Unlike renting services consumers here should have longer contracts. In case of Lütteleihen, the minimum time period of a contract is 2-months. "Besides the traditional second hand shops, there are many online second hand markets such as "Walk in my Closet" where consumers can buy second hand luxury shoes, clothing, bags, or accessories from a commercial website" (Iran and Schrader 2017: 475).

Nowadays, there are plenty of the existing practices of CFC in societies. The examples that are mentioned in this section and other similar ones that are continuously becoming more and more commonplace in societies can bring positive environmental and social impacts.

8 Summary and Outlook

In this chapter, the author summarized some of the most important concepts related to sustainable fashion. An emphasis has been put on the consumption phase and specifically on the alternative and innovative fashion initiatives.

Systematically, the efforts towards fostering sustainability in fashion have been categorized into three groups: governance, production and consumption. First of all, we need a proper legislation to promote and control sustainability in the fashion industry. These rules and their implementation should be continuously monitored not only by governmental agencies but also by NGOs and independent media. Then in the production phase, efforts should be put on reaching the maximum resource efficiency according to the principles of sustainable development. Material-, intellectual-, and human resources are discussed in this group.

The sustainability of fashion is not only determined by sustainable governance and production but also by consumers, their intentions, behaviors, and habits. Consumers play a significant role in decreasing the harmful effects of fashion consumption on the environment. As it is classified in this chapter these effects occur in the purchase, the usage, and the post-usage phase of fashion consumption. Initially, consumers make a decision on buying new bio/fairtrade products or using second hand ones. Afterwards, consumers make a decision on how they want to use and maintain their fashion products. Finally, they need to determine what should be done with the garments after the end of the usage period. In the final phase, they can decide to reuse the garments by selecting one of the opportunities that CFC offers.

Buying less, wearing long, maintaining well, and proper feeding back into the system for recycling after use are some examples of sustainable fashion consumption (Allwood et al. 2006). Consumers can also decide to use second hand garments instead of buying new ones. They can decide to return their used clothes to the life cycle by making them available to the second hand market. If consumers decide to use second hand clothing either by swapping, borrowing, lending through Peer-to-Peer CFC, or by renting, leasing, or buying second hand clothes from companies, they enter the area of CFC with all of its ownership-less opportunities. There is a good potential of achieving environmental and social benefits by extending CFC in the fashion industry. Having more comprehensive studies on different types of CFC in combination with generating good strategies could improve sustainability in the fashion industry. There is still a lack of academic research on the opportunities and barriers of different types of CFC, motives of the current consumers who are participating in CFC, as well as best practices and governance structure of CFC practices. In addition, a comprehensive life cycle assessment could confirm the positive environmental impact of CFC and could reveal the amount of environmental opportunities that CFC can offer.

References

Allwood JM, Laursen SE, Rodríguez CM, Bocken NMP (2006) Well dressed?: the present and future sustainability of clothing and textiles in the United Kingdom. University of Cambridge, Institute for Manufacturing, Cambridge

Armstrong CM, Niinimäki K, Kujala S, Karell E, Lang C (2015) Sustainable product-service systems for clothing: exploring consumer perceptions of consumption alternatives in Finland. J Clean Prod 97:30–39. https://doi.org/10.1016/j.jclepro.2014.01.046

Auger P, Devinney TM, Louviere JJ, Burke PF (2010) The importance of social product attributes in consumer purchasing decisions: a multi-country comparative study. Int Bus Rev 19 (2):140–159. https://doi.org/10.1016/j.ibusrev.2009.10.002

Baden S, Barber C (2005) The impact of the second hand clothing trade on developing economies. http://oxfamilibrary.openrepository.com/oxfam/bitstream/10546/112464/1/rr-impact-second-hand-clothing-trade-developing-countries-010905-en.pdf. Accessed 18 Aug 2017

Balderjahn I, Buerke A, Kirchgeorg M, Peyer M, Seegebarth B, Wiedmann K-P (2013) Consciousness for sustainable consumption: scale development and new insights in the economic dimension of consumers' sustainability. AMS Rev 3(4):181–192. https://doi.org/10.1007/s13162-013-0057-6

Baldwin C, von Hippel E (2011) Modeling a paradigm shift: from producer innovation to user and open collaborative innovation. Organ Sci 22(6):1399–1417. https://doi.org/10.1287/orsc.1100.0618

Bartlett C, McGill I, Willis P (2013) Textiles flow and market textiles flow and market development opportunities in the UK. http://www.wrap.org.uk/system/files/private/Textile%20Flows%20and%20Market%20Development%20Opportunities%20Updated%20June%202013v2.pdf. Accessed 18 Aug 2017

Belz F-M (2006) Marketing in the 21st century. Bus Strateg Environ 15(3):139–144. https://doi.org/10.1002/bse.529
Bendell J, Kleanthous A (2007) Deeper luxury: quality and style when the world matters. http://assets.wwf.org.uk/downloads/luxury_report.pdf. Accessed 18 Aug 2017
Birtwistle G, Moore CM (2006) Fashion adoption in the UK: a replication study. http://www.anzmac.org/conference_archive/2006/documents/Birtwhistle_Grete.pdf. Accessed 18 Aug 2017
Birtwistle G, Moore CM (2007) Fashion clothing – where does it all end up? Int J Retail Distrib Manag 35(3):210–216. https://doi.org/10.1108/09590550710735068
Black S (2008) Eco-chic: the fashion paradox. Black Dog, London
Black S, Anderson S (2010) Making sustainability fashionable: profile of the Danish fashion company noir. Fash Pract J Des Creat Process Fash Ind 2(1):121–128. https://doi.org/10.2752/175693810X12640026716555
Botsman R, Rogers R (2010) What's mine is yours: the rise of collaborative consumption, 1st edn. Harper Business, New York
Botsman R, Rogers R (2011) What's mine is yours: how collaborative consumption is changing the way we live, Revised and updated edn. Collins, London
BSR (2009) Apparel industry life cycle carbon mapping. https://www.bsr.org/reports/BSR_Apparel_Supply_Chain_Carbon_Report.pdf. Accessed 18 Aug 2017
Butler S (2007) M&S to spend £200m on green strategy. https://www.thetimes.co.uk/article/mands-to-spend-pound200m-on-green-strategy-sljlmthw3mp. Accessed 17 Apr 2017
Carter CR, Rogers DS (2008) A framework of sustainable supply chain management: moving toward new theory. Int J Phys Distrib Logist Manag 38(5):360–387. https://doi.org/10.1108/09600030810882816
Claudio L (2007) Waste couture: environmental impact of the clothing industry. Environ Health Perspect 115(9):A449–A454
DEFRA (2008) Sustainable clothing roadmap briefing note December 2007: sustainability impacts of clothing and current interventions. DEFRA, London
Dickson MA (2001) Utility of no sweat labels for apparel consumers: profiling label users and predicting their purchases. J Consum Aff 35(1):96–119. https://doi.org/10.1111/j.1745-6606.2001.tb00104.x
Donaldson T (2016) World's top 10 Garment exporters rated high risk for modern slavery. https://sourcingjournalonline.com/worlds-top-10-garment-exporters-rated-high-risk-modern-slavery/. Accessed 18 Aug 2017
Eder-Hansen J, Kryger J, Morris J, Sisco C, Watson D, Kiørboe N, Petersen SD, Larsen KB, Burchardi I (2012) The nice consumer: toward a framework for sustainable fashion consumption in the EU. http://nordicfashionassociation.com/sites/default/files/bsr_nice_consumer_discussion_paper_0.pdf. Accessed 18 Aug 2017
Faisal MN (2010) Sustainable supply chains: a study of interaction among the enablers. Bus Process Manag J 16(3):508–529. https://doi.org/10.1108/14637151011049476
Fineman S (2001) Fashioning the environment. Organization 8(1):17–31. https://doi.org/10.1177/135050840181002
Fletcher K (2008) Sustainable fashion and textiles: design journeys. Earthscan, London
Füller J, Matzler K, Hutter K, Hautz J (2012) Consumers' creative talent: which characteristics qualify consumers for open innovation projects? An exploration of asymmetrical effects. Creat Innov Manag 21(3):247–262. https://doi.org/10.1111/j.1467-8691.2012.00650.x
Gata JE (2015) The sharing economy, competition and regulation. https://www.competitionpolicyinternational.com/assets/Europe-Column-November-Full.pdf. Accessed 17 Apr 2017
Greenpeace (2015) Usage & attitude mode/Fast fashion: Ergebnisbericht. https://www.greenpeace.de/files/publications/20151123_greenpeace_detox_ergebnisbericht_fast_fashion.pdf. Accessed 18 Aug 2017

Guardian (2001) Story of the blues. https://www.theguardian.com/g2/story/0,,497788,00.html. Accessed 18 Aug 2017
Harrison R, Newholm T, Shaw D (2005) The ethical consumer. Sage, London
Hawley JM (2006) Digging for diamonds: a conceptual framework for understanding reclaimed textile products. Cloth Text Res J 24(3):262–275. https://doi.org/10.1177/0887302X06294626
Hirscher AL, Iran S (2017) Sewing cafe- micro-utopia for sustainable fashion production and consumption? Working Paper
ILO (2005) Promoting fair globalization in textiles and clothing in a post-MFA environment: report for discussion at the Tripartite meeting on promoting fair globalization in textiles and clothing in a post-MFA environment. http://www.ilo.org/wcmsp5/groups/public/---ed_dialogue/---sector/documents/meetingdocument/wcms_161673.pdf. Accessed 18 Aug 2017
ILO (2014) Wages and working hours in the textiles, clothing, leather and footwear industries. http://www.ilo.org/wcmsp5/groups/public/@ed_dialogue/@sector/documents/publication/wcms_300463.pdf. Accessed 18 Aug 2017
Iran S, Schrader U (2017) Collaborative fashion consumption and its environmental effects. J Fash Mark Manag 21(4):468–482. https://doi.org/10.1108/JFMM-09-2016-0086
Jastram S (2007) Corporate social responsibility im interkulturellen Raum. Entwicklungsethnologie 16(2):73–83
Jegethesan K, Sneddon JN, Soutar GN (2012) Young Australian consumers' preferences for fashion apparel attributes. J Fash Mark Manag 16(3):275–289. https://doi.org/10.1108/13612021211246044
Joergens C (2006) Ethical fashion: myth or future trend? J Fash Mark Manag 10(3):360–371. https://doi.org/10.1108/13612020610679321
Kim H-S, Damhorst ML (1998) Environmental concern and apparel consumption. Cloth Text Res J 16(3):126–133. https://doi.org/10.1177/0887302X9801600303
Laursen SE, Hansen J, Bagh J, Jensen OK, Werther I (1997) Environmental assessment of textiles. http://www2.mst.dk/Udgiv/publications/1997/87-7810-838-1/pdf/87-7810-838-1.pdf. Accessed 30 Aug 2017
Lewis H, Gertsakis J (2001) Design + environment: a global guide to designing greener goods. Greenleaf, Sheffield
Lo CK, Yeung AC, Cheng T (2012) The impact of environmental management systems on financial performance in fashion and textiles industries. Int J Prod Econ 135(2):561–567. https://doi.org/10.1016/j.ijpe.2011.05.010
McDonough W, Braungart M (2010) Cradle to cradle remaking the way we make things. MacMillan, New York
Meyer A (2001) What's in it for the customers? Successfully marketing green clothes. Bus Strateg Environ 10(5):317–330. https://doi.org/10.1002/bse.302
Mhango MW, Niehm LS (2005) The second-hand clothing distribution channel. J Fash Mark Manag 9(3):342–356. https://doi.org/10.1108/13612020510610462
Morgan LR, Birtwistle G (2009) An investigation of young fashion consumers' disposal habits. Int J Consum Stud 33(2):190–198. https://doi.org/10.1111/j.1470-6431.2009.00756.x
Moulds J Child labour in the fashion supply chain: where, why and what can be done. https://labs.theguardian.com/unicef-child-labour/. Accessed 28 Aug 2017
Nagurney A, Yu M (2012) Sustainable fashion supply chain management under oligopolistic competition and brand differentiation. Int J Prod Econ 135(2):532–540. https://doi.org/10.1016/j.ijpe.2011.02.015
Niinimäki K (2009) Consumer values and ecofashion in the future. In: Koskela M, Vinnari M (eds) Proceedings of the conference: future of the consumer society. Finland Futures Research Centre, Turku, Finland, pp 125–134
Niinimäki K (2010) Eco-clothing, consumer identity and ideology. Sustain Dev 18(3):150–162. https://doi.org/10.1002/sd.455
NIRI (2006) Recycling of low grade clothing waste. http://www.oakdenehollins.com/pdf/defr01_058_low_grade_clothing-public_v2.pdf

Ochoa LMC (2010) Will 'eco-fashion' take off? A survey of potential customers of organic cotton clothes in London. AD-minister 16:118–131
OTA (2015) Organic cotton facts. https://ota.com/sites/default/files/indexed_files/OrganicCottonFacts2015_0.pdf. Accessed 28 Aug 2017
Palmer A, Clark H (2005) Old clothes, new looks: second hand fashion. Dress, body, culture, 1360-466X. Berg, Oxford
Paulins VA, Hillery JL (2009) Ethics in the fashion industry. Fairchild Books, New York, NY
Pears KE (2006) Fashion re-consumption: developing a sustainable fashion consumption practice influenced by sustainability and consumption theory. Master's dissertation, Architecture and Design
Pedersen ERG, Netter S (2015) Collaborative consumption: business model opportunities and barriers for fashion libraries. J Fash Mark Manag 19(3):258–273. https://doi.org/10.1108/JFMM-05-2013-0073
Rinaldi FR, Testa S (2015) The responsible fashion company: integrating ethics and aesthetics in the value chain. Greenleaf, Sheffield
Ritzer G, Jurgenson N (2010) Production, consumption, prosumption: the nature of capitalism in the age of the digital 'prosumer'. J Consum Cult 10(1):13–36. https://doi.org/10.1177/1469540509354673
Rivoli P (2015) The travels of a T-shirt in the global economy: an economist examines the markets, power, and politics of world trade, second edition, new preface and epilogue, with updates on economic issues and main characters. Wiley, New Jersey
Rudell F (2006) Shopping with a social conscience: consumer attitudes toward sweatshop labor. Cloth Text Res J 24(4):282–296. https://doi.org/10.1177/0887302X06293063
Sampson LK (2009) Consumer analysis of purchasing behavior for green apparel. Master's dissertation, North Carolina State University
Sayer K (2007) Slow food, slow fashion. http://www.treehugger.com/style/slow-food-slow-fashion.html. Accessed 17 Apr 2017
Schaefer A, Crane A (2005) Addressing sustainability and consumption. J Macromark 25(1):76–92. https://doi.org/10.1177/0276146705274987
Schor JB, Fitzmaurice CJ (2015) Collaborating and connecting: the emergence of the sharing economy. In: Reisch L, Thøgersen J (eds) Handbook of research on sustainable consumption. Edward Elgar, Cheltenham, pp 410–425
Schrader U (2001) Konsumentenakzeptanz eigentumsersetzender Dienstleistungen: Konzeption und empirische Analyse. Markt und Konsum 10. Lang, Frankfurt am Main [u.a.]
Schrader U (2007) The moral responsibility of consumers as citizens. Int J Innov Sustain Dev 2 (1):79. https://doi.org/10.1504/IJISD.2007.016059
Schultz MF (ed) (2010) Sustainable fashion: issues to be addressed. Kolding School of Design, Laboratory for Design, Innovation and Sustainability, Kolding
Shaw D, Hogg G, Wilson E, Shiu E, Hassan L (2006) Fashion victim: the impact of fair trade concerns on clothing choice. J Strateg Mark 14(4):427–440. https://doi.org/10.1080/09652540600956426
Stanszus L, Iran S (2015) Sustainable fashion. In: Idowu SO, Capaldi N, Fifka M, Zu L, Schmidpeter R (eds) Dictionary of corporate social responsibility. Springer, Cham
Thøgersen J, Crompton T (2009) Simple and painless? The limitations of spillover in environmental campaigning. J Consum Policy 32(2):141–163. https://doi.org/10.1007/s10603-009-9101-1
Tischner U, Charter M (2001) Sustainable product design. In: Charter M, Tischner U (eds) Sustainable solutions: developing products and services for the future. Greenleaf, Sheffield, pp 118–139
Tukker A, Charter M, Vezzoli C, Stø E, Munch Andersen M (eds) (2008) Perspectives on radical changes to sustainable consumption and production, System innovation for sustainability, vol 1. Greenleaf, Sheffield

US EPA (2014) Advancing sustainable materials management: facts and figures. https://www.epa.gov/smm/advancing-sustainable-materials-management-facts-and-figures. Accessed 28 Aug 2017

Williams S, Taylor J, Howards M (2005) The ethical consumerism report 2005. The Cooperative Bank. http://b.3cdn.net/nefoundation/155ba657edf544c402_69m6bpbos.pdf. Accessed 30 Aug 2017

Woolridge AC, Ward GD, Phillips PS, Collins M, Gandy S (2006) Life cycle assessment for reuse/recycling of donated waste textiles compared to use of virgin material: an UK energy saving perspective. Resour Conserv Recycl 46(1):94–103. https://doi.org/10.1016/j.resconrec.2005.06.006

WRAP (2012) Valuing our clothes: the true cost of how we design, use and dispose of clothing in the UK. http://www.wrap.org.uk/sites/files/wrap/VoC%20FINAL%20online%202012%2007%2011.pdf. Accessed 31 May 2016

Zafarmand SJ, Sugiyama K, Watanabe M (2003) Aesthetic and sustainability: the aesthetic attributes promoting product sustainability. J Sustain Prod Des 3(3–4):173–186. https://doi.org/10.1007/s10970-005-6157-0

Part IV
Conclusion

New Business and Governance Approaches to Sustainable Fashion: Learning from the Experts

Sarah Margaretha Jastram and Anna-Maria Schneider

This edited volume comprises articles by managers, scientists and political experts in the field of sustainable fashion. They share their best practice experience relating to the joint goal of a more sustainable, humanistic, and responsible fashion industry. We find an exciting variety of approaches in this book that in sum represent a rich pool of insights and cases for anyone interested in the study of new management perspectives and in the systematic advancement of sustainable fashion.

At the core of many leading business approaches to sustainability is circular thinking, meaning that the business process does not only involve the design, production, marketing and distribution of fashion items but also its sustainable recycling, repair, reuse or resale.

These approaches are greatly beneficial for the environment as they reduce the massive piles of rubbish produced by the fast fashion industry. Circular business models also support and serve consumers who wish to simplify their lives and their wardrobes, consumers who are unwilling to live a lifestyle of increasing consumption and waste creation, and those who prefer the selection and ownership of long lasting fashion items over less durable low-quality clothes.

Filippa K's circular approach to sustainable fashion and its membership in the Fair Wear Foundation demonstrate an ongoing strong commitment in this context. Another impressive example is the innovative, young brand MELAWEAR which invented great sustainable fashion following the Cradle to Cradle principles of recycling and strictly working with Fairtrade and GOTS certified materials. TAUKO's upcycling approach as well as Samira Iran's chapter on different forms

S. M. Jastram (✉)
Department of Strategy and Leadership, Hamburg School of Business Administration,
Dr. Jürgen Meyer Chair of International Business Ethics and Sustainability, Hamburg, Germany
e-mail: sarah.jastram@hsba.de

A.-M. Schneider
Federal Ministry for Economic Cooperation and Development, Berlin, Germany

S. M. Jastram, A.-M. Schneider (eds.), *Sustainable Fashion*, Management for Professionals, https://doi.org/10.1007/978-3-319-74367-7_14

of collaborative fashion consumption show further evidence of the continuing establishment of circular business models in the fashion industry. A central question of the future is how to bring the idea of circular economy to the mass market and how to scale up individual approaches to sector-wide engagement.

A second core topic of this volume is the supply chain and the fundamental question of how decent and human working conditions can be realized including living wages and effective workers' interest representation. While audit-oriented control-based forms of supply chain governance used to be prevailing industry approaches, we can now observe a trend towards more collaboration oriented forms of supply chain governance. This does not mean that audits do not play a role anymore in today's business approaches to sustainable supply chains, but there is a need to go beyond auditing and control-based forms of governance (Jastram and Schneider 2015). Front runner firms like Tchibo for instance are increasingly pursuing more bottom-up and inclusive forms of governance as illustrated in the chapter authored by Nanda Bergstein. H&M is another example of a brand that is intensely working on new methods of supply chain governance as the firm remains under strong stakeholder pressure. In this book, Hendrik Heuermann explains the importance of collective bargaining in the context of the ongoing debate about best ways to support and to enable living wages in textile factories in developing countries.

The assessment of actual impacts of any sustainable governance initiative is challenging and related methods are currently intensely debated in the academic literature (Baumann-Pauly and Jastram 2018). In this context, Otto is introducing an interesting example of a value chain impact measurement in this book.

From a stakeholder, especially from a consumer's perspective, transparency and traceability are core conditions for functioning market mechanisms supporting sustainable fashion (Berberyan et al. 2018). The availability of relevant information on product characteristics and production conditions support enlightened consumer voting for product and brands with the best sustainability credentials as well as the identification of those brands which are not yet sufficiently integrating sustainability measures into their core business operations. Isabel Laura Ebert and Robin Cornelius demonstrate this in this book.

Besides these market and business driven initiatives, governments have increasingly been putting international supply chains into the focus of international meetings and regulative activities. A core governance framework in this context are the UN Guiding Principles for Business and Human Rights developed by John Ruggie and explained by Natascha Weisert in this book. The heads of state and government of the Group of Seven (G7) have further raised attention to the topic of sustainable supply chains in the fashion sector by placing it among the core topics of their leaders' declaration in Elmau 2015. Roger A. Fischer explains the underlying governance processes and mechanisms as well as its challenges.

While business practice as well as state governance both seem to be promising approaches to today's challenges in international value creation processes, neither of the two is currently sufficient to solve all relating social and environmental issues alone. Accordingly, Baumann-Pauly, Labowitz and Stein argue for a transformative industry approach based on the shared responsibility model.

In this context, the notion of a 'smart mix' has recently gained attention in the international political arena as well as on business round tables. The exact format of an ideal combination of different elements and instruments of governance has not yet been agreed upon. However, as this volume shows, most instruments, initiatives, processes, and tools are interrelated and interdependent in terms of their potential impact (see graphical illustration).

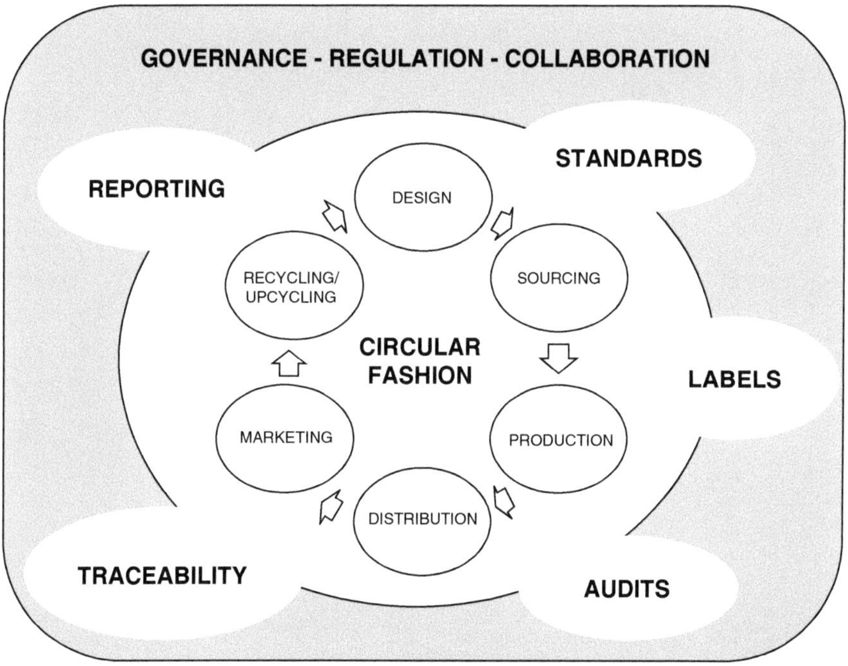

Sustainable fashion cannot be realized by just one actor or by using just one business tool or governance instrument. The transformation of the sector requires concerted action of businesses, governments, NGOs, and consumers to clearly demonstrate a coalition for change and a willingness and power to end inhumane and unsustainable business practices in the twenty-first century.

The necessary mutual learning process is ongoing and will continue to involve trial as well as errors. Even though sustainable fashion management and governance have progressed strongly during the last years, human right violations as well as environmental pollution still exist in daily business of some players in the fashion industry. Yet, the momentum for change is strong and this edited volume will hopefully support this important mission.

References

Baumann-Pauly D, Jastram SM (2018) Assessing human rights issues in the fashion industry – challenges for investors. In: Radi Y (ed) Research handbook on human rights and investments. Edward Elgar (in press)

Berberyan Z, Jastram SM, Friedman B (2018) Drivers and obstacles of ethical fashion consumption. In: Becker-Leifhold C, Heuer M (eds) Eco friendly and fair: fast fashion and consumer behavior. Greenleaf, Sheffield (in press)

Jastram S, Schneider AM (2015) Sustainable fashion governance at the example of the partnership for sustainable textiles. Umweltwirtschaftsforum 23(4):205–212

Erratum to: Sustainable Fashion

Sarah Margaretha Jastram and Anna-Maria Schneider

Erratum to:
S. M. Jastram, A.-M. Schneider (eds.), *Sustainable Fashion*, Management for Professionals, https://doi.org/10.1007/978-3-319-74367-7

This book was inadvertently published with the incorrect editor sequence as Anna-Maria Schneider and Sarah Margaretha Jastram. The editor sequence has been corrected throughout the book except chapter 1 as Sarah Margaretha Jastram and Anna-Maria Schneider.

The updated online version of this book can be found at https://doi.org/10.1007/978-3-319-74367-7

S. M. Jastram, A.-M. Schneider (eds.), *Sustainable Fashion*, Management for Professionals, https://doi.org/10.1007/978-3-319-74367-7_15

MIX
Papier aus verantwortungsvollen Quellen
Paper from responsible sources
FSC® C105338

If you have any concerns about our products,
you can contact us on
ProductSafety@springernature.com

In case Publisher is established outside the EU,
the EU authorized representative is:
Springer Nature Customer Service Center GmbH
Europaplatz 3, 69115 Heidelberg, Germany

Printed by Libri Plureos GmbH
in Hamburg, Germany